Spoof Plasmons

Synthesis Lectures on Electromagnetics

Editor
Aklesh Lakhtakia, *The Pennsylvania State University*

Spoof Plasmons
Tatjana Gric
2020

The Transfer-Matrix Method in Electromagnetics and Optics
Tom G. Mackay and Akhlesh Lakhtakia
2020

Spoof Plasmons

Tatjana Gric

ISBN: 978-3-031-00895-5   paperback
ISBN: 978-3-031-02023-0   ebook
ISBN: 978-3-031-00138-3   hardcover

DOI 10.1007/978-3-031-00138-3

A Publication in the Springer series
SYNTHESIS LECTURES ON ELECTROMAGNETICS

Lecture #2
Series Editor: Aklesh Lakhtakia, *The Pennsylvania State University*
Series ISSN
ISSN pending.

Spoof Plasmons

Tatjana Gric
Vilnius Gediminas Technical University

SYNTHESIS LECTURES ON ELECTROMAGNETICS #2

ABSTRACT

The fundamental optical excitations that are confined to a metal/dielectric interface are the surface plasmon polaritons (SPPs), as described by Ritchie. SPPs can be referred to as electromagnetic excitations existing at an interface between two media, of which at least one is conducting. Investigating spoof plasmons in a semiconductor is becoming an increasingly active area of research. The field of plasmonics deals with the application of surface and interface plasmons. It is an area in which surface plasmon-based circuits merge the fields of photonics and electronics at the nanoscale. Recently, an idea of engineering surface plasmons at lower frequencies was suggested. It was concluded in that the existence of holes in the structure can lower the frequency of existing surface plasmons. Thus, by cutting holes or grooves in metal surfaces, it is possible to take concepts such as highly localized waveguiding and superfocusing to lower frequencies, particularly to the THz regime, where plasmonics could enable near-field imaging and biosensing with unprecedented sensitivity. It is the main reason to use the terminology "spoof surface plasmons" for the bound surface waves propagating along the perforated structures. The book's title *Spoof Plasmons* demonstrates that it is devoted to exhibiting the current state of the art of the dynamic and vibrant field of photonic metamaterials reaching across various disciplines, suggesting exciting applications in chemistry, material science, biology, medicine, and engineering.

KEYWORDS

surface plasmon polaritons, plasmonics, photonics, near-field imaging, biosensing, material science, engineering

Contents

Preface

Surface Plasmons (SP), or Surface Plasmon Polaritons (SPP), is a designation given to the quasi-particles representing the collective oscillation of the surface charges at a metal/dielectric interface. From the material point of view, the necessary condition for the excitation of the SPPs modes is the existence of a 2D electronic gas formed at the interface between a conductor with complex permittivity and a dielectric with positive permittivity. Collective oscillations of surface charges demonstrate themselves in the form of longitudinal surface waves that propagate along the metals/dielectric interface. The fundamental optical excitations that are confined to a metal/dielectric interface are the SPPs, as described by Ritchie [1]. SPPs can be referred to as electromagnetic excitations existing at an interface between two media, of which at least one is conducting [2]. The important properties exhibited by plasmonic elements like silver, gold, copper, and aluminum are surface plasmons which concentrate the optical energy in nanoscale [3, 4]. The existent plasmonic properties (like surface plasmon mode) in materials are dramatically affected by their dielectric constants. Surface plasmon mode is an important property. In doing so, the optical energy in nanoscale domain [5, 6] can be concentrated. It is worthwhile mentioning that the dielectric constants are the complex quantities. The real part is the reflection and imaginary part represents the absorption or loss. Therefore, on the basis of dielectric constant value it is possible to detect good and bad plasmonic materials and also to decide if the material possesses plasmonic properties or not. Those materials are the plasmonic materials for which dielectric constant ε_m has a negative real part, Re $\varepsilon_m < 0$ and imaginary part of dielectric constant is much less than the real part of dielectric constant (Im $\varepsilon_m <<$ Re ε_m). Under these two conditions surface plasmon resonances are most effective due to minimum losses in metals. Silver and gold exhibit plasmon properties in visible range because the above mentioned two properties are satisfied in visible range [7, 8].

Utilizing the terahertz range for better and faster communications [9], sensing [10], medicine [11], and security [12] has created a need for devices capable of operating in the desired frequency range of 0.1–30 THz. One of the principles that can serve as the basis of guiding, coupling, and modulating THz waves is the surface plasmon—an interface wave propagating at the boundary of negative (conductive) and positive (dielectric) permittivity material. Traditional plasmonic materials usable in the invisible/near infrared range, such as noble metals, are unsuitable for uses in the THz regime. The main their drawback is the low confinement to the metal. Thus, the wave is weakly bound to the interface. The described phenomenon is sometimes called the Zenneck plasmon [13]. Metallic properties of semiconductors, which possess special carrier levels, are shifted to lower frequencies—microwave, terahertz, and far infrared. They therefore can be employed as the building blocks for THz devices. Furthermore, they al-low for much-needed control of their electromagnetic properties. In the manufacturing one can adjust the carrier levels by doping and after the manufacturing the properties can be engineered by light [14], temperature [15], electric gating [16], and external magnetic field [17].

Investigations of light-sensitive media designed for recording volume (thick) holograms are now performed in a typical unified manner: formation of a test light field (action); obtaining a material response of the medium to this action; studying deviations of this response from the test response; and establishing relationships between the observed deviations and mechanisms of photoinduced transformations of matter. As the test light field is three-dimensional and is formed in space due to interference of two intersecting plane light waves, the material responses of the medium placed into such a field and the objects of the study are thick gratings of the transmission or reflection types. Detection of spatial deviations of these gratings from the "light" test gratings satisfying the Bragg condition is most often performed through comparisons (or fitting) of their angular or, in other words, selective, diffraction and transmission characteristics. The research technique corresponding to this approach was developed in [18] and other investigations based on Kogelnik's theory of thick holographic gratings [19].

Investigations of spoof plasmons in a semiconductor is becoming an increasingly active area of research. Actually, the presence of surface waves was originally proven either for dielectric materials placed over a metal surface or for the metallic surface with a periodic repetition of obstacles of holes in the direction of propagation [20–22]. However, these surface waves can also be excited replacing the metallic surface with the semiconductor one [23]. Spoof plasmons are bound electromagnetic (EM) waves at frequencies outside the plasmonic range mimicking ("spoofing") SPPs, which propagate on periodically corrugated metal surfaces [24].

The field of plasmonics deals with the application of surface and interface plasmons. It is an area in which surface-plasmon based circuits merge the fields of photonics and electronics at the nanoscale [25]. Surface-plasmon photonics can open broad avenues for constructing nanoscale photonic circuits that will be able to carry optical signals and electric currents [26, 27]. Surface plasmons can also be employed in the design, fabrication, and characterization of subwavelength waveguide components [28–41]. Modulators and switches have also been investigated [42, 43] in the framework of plasmonics. The use of surface plasmons as mediators in the transfer of energy from donor to acceptors molecules on opposite sides of metal films [39] is also promising. Among various surface plasmon polaritons waveguides [40], the metal–insulator–metal (MIM) waveguides have been extensively investigated [41–43]. Besides that, they have attracted great interest from researchers, because the highly confined surface plasmon modes in the insulator region can propagate in a sharp bend with low additional loss [44] and they are easily manufactured using existing nanofabrication techniques [45]. A large stream of papers is dedicated to the studies of nanostructured surfaces in the context of fluctuation-induced electromagnetic forces and energy transfer [46–50]. These systems are introduced as the cavities designed between a planar metallic mirror and a nanostructured grating (or between two nanostructured gratings) separated by vacuum gaps or dielectric material.

Recently, an idea of engineering surface plasmons at lower frequencies was suggested. It was concluded in [51] that the existence of holes in the structure can lower the frequency of existing surface plasmons. Thus, by cutting holes or grooves in metal surfaces it is possible to take concepts such as highly localized waveguiding [30, 34] and superfocusing [40, 52] to lower frequencies, particularly to the THz regime [53] where plasmonics could enable near-field imaging and biosensing [54] with unprecedented sensitivity.

In recent years, electromagnetic waves propagating at an interface between a metal and a dielectric have been of significant interest. An experimental study of propagation of a THz Zenneck surface wave on an aluminium sheet is presented in [13]. The properties of long-range SPPs are reviewed in [55]. It is shown in [56] that a simple waveguide, namely a bare metal wire, can be used to transport terahertz pulses with virtually no dispersion, low attenuation, and with a remarkable structural simplicity.

Although most plasmonic research so far has focused on the near-infrared and optical ranges of the electromagnetic spectrum (where noble metals support highly confined surface waves), there exists an increasing interest in transferring SPPs-based photonics to lower frequencies, such as microwave, millimeter, or sub-millimeter waves. It is demonstrated in [57] that a one-dimensional (1D) periodically corrugated metal film can be used to create planar terahertz waveguides. Based on a modal expansion of electromagnetic fields, a rigorous method for analyzing SPPs on a periodically corrugated metal surface has been formulated in [58]. It is shown in [59] that the dispersion relation of SPPs propagating along a perfectly conducting wire can be tailored by corrugating its surface with a periodic array of radial grooves.

However, in these spectral ranges, noble metals behave like perfect electric conductors, whose surface charges are able to screen any external EM excitation with extreme efficiency, preventing the formation of tightly bound SPPs [60]. It has been shown in [61–63] that binding of EM fields to a metal surface can be increased by its corrugation. Highly confined spoof surface plasmons were theoretically predicted to exist in a perforated metal film coated with a thin dielectric layer [64]. A new type of a waveguide scheme for terahertz circuitry based on the concept of spoof surface plasmons is presented in [65]. This structure is composed of a 1D array of L-shaped metallic elements horizontally attached to a metal surface. The dispersion relation of the surface electromagnetic modes supported by this system has a very weak dependence on the lateral dimension, and the modes are very deep subwavelength confined with a long-enough propagation length.

The question of localization of spoof plasmons is still open. The spoof plasmon behavior has been considered in [1]. The authors have investigated a surface of a metal perforated with a 1D periodic array of rectangular grooves. Another question that remains open is the calculation of the effective permittivities for triangular grooves. A number of reported results describing the calculation of the effective dielectric constants for the grooved surfaces is limited. The effective dielectric constants for the rectangular grooves in a perfect conductor were obtained in [36].

The experimental evidence of the resonance of a surface mode propagating across the surface of a near perfectly conducting substrate perforated with holes has been presented in [13]. Up until now, the spoof SPPs at THz frequencies on corrugated semiconductor surfaces have not been studied in detail, particularly investigating the losses of these spoof SPPs. The analysis of the previous theoretical studies dedicated to spoof SPPs shows that the researchers have been dealing with metals as with a perfect conductor. In this case, the spoof SPPs are lossless [51, 60–64]. Although some papers deal with spoof plasmon losses [20], the question of how losses affect propagation of spoof plasmons has not been studied in detail yet. Moreover, the propagation of spoof SPPs on corrugated semiconductor surfaces is of special interest. To the best of our knowledge, this topic has not been covered yet. In light of this, the study of spoof SPPs at THz frequencies is of interest for both basic physics and possible applications in device design.

This book is devoted to exhibiting the current state of the art of the dynamic and vibrant field of photonic metamaterials reaching across various disciplines, suggesting exciting applications in chemistry, material science, biology, medicine, and engineering. It will illuminate recent advances in the wider photonic metamaterials field, such as (to mention a few) active metamaterials and metasurfaces, self-organized nanoplasmonic metamaterials, graphene metamaterials, metamaterials with negative or vanishing refractive index, and topological metamaterials facilitating ultraslow broadband waves on the nanoscale and novel applications, such as stopped-light lasing. Also, we focus on the connection of the two previous topics: metamaterials and metasurfaces. Nevertheless, most of the aforementioned plasmonic waveguides and devices incorporate metals. Recently, it has been demonstrated that SPPs can also be confined and propagate on corrugated semiconductor strips [23]. The former issue opens the wide avenues to the design of low-loss propagation for SPPs. In this study, we have numerically investigated the properties of SPP on the corrugated SIS waveguides made of the semiconductor and transparent conducting oxides (TCO).

Tatjana Gric
September 2020

CHAPTER 1

Methods

1.1 PROPAGATION OF PLASMONS IN SEMICONDUCTORS

Consider a corrugated semiconductor surface in which an array of grooves exists. This structure is shown in Fig. 1.1. The surface is continuous in the x-direction and there are two different materials in the $z-y$ plane: the semiconductor occupies the lower region and above there is a dielectric as a surrounding medium. Such an interface can be modeled as a three-layer structure, consisting of a homogeneous anisotropic layer of a thickness h—see Fig. 1.1—describing corrugations, placed between a semiconductor and a dielectric. The approximation of the central layer as an anisotropic effective medium is possible as long as the period d is much smaller than the wavelength of the electromagnetic field.

In order to find the dielectric parameters of the effective medium, consider a periodic assembly of parallel plates. The effective dielectric constants of such an assembly are as follows [68]:

$$\varepsilon_x = \varepsilon_z = \frac{(d-a)\varepsilon_s + a\varepsilon_g}{d} \tag{1.1}$$

$$\varepsilon_y = \frac{d}{(d-a)/\varepsilon_s + a/\varepsilon_g}. \tag{1.2}$$

Here, ε_d is the permittivity of the surrounding media, ε_g – is the permittivity of the material, filling the groves, ε_s – is the permittivity of the semiconductor defined by the Drude model, i.e., $\varepsilon(\omega) = \varepsilon_\infty \left(1 - \frac{\omega_p^2}{\omega^2}\right)$, ε_∞ is the background permittivity, and ω_p is the plasma frequency.

It is interesting to notice that heavily doped silicon ($n > 2.2 \times 10^{19}$ cm^{-3}) has been shown to exhibit metallic properties at terahertz frequencies [74, 75] and has the potential to replace metals in such applications [76]. In order to consistently assess the effect of loss on the waves with the Drude permittivity it is suggested first to review the major features of the waves in respective lossless case.

Using the effective medium approximation derived in the previous section, one can describe the transverse magnetic spoof plasmon mode propagating in the structure. The only nonzero components of the magnetic and electric fields expressed in terms of $\varepsilon_x, \varepsilon_y, \varepsilon_z$ are $H_z, E_x,$

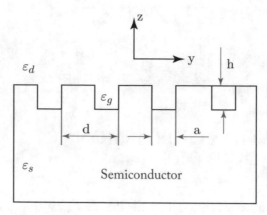

Figure 1.1: Geometry of structured semiconductor surface.

E_y [24]:

$$H_x = Ae^{-jky} \begin{cases} e^{-\kappa z}, & z > 0 \\ \dfrac{\cos\left(k_g(z+h)\right)}{\cos\left(k_g h\right)}, & -h < z < 0 \end{cases} \tag{1.3}$$

$$E_y = -\frac{\partial_z H_x}{ik_0 \varepsilon_y} = -\frac{Ai e^{iky}}{k_0} \begin{cases} \dfrac{\kappa}{\varepsilon_d} e^{-\kappa z}, & z > 0 \\ \dfrac{k_g}{\varepsilon_y} \cdot \dfrac{\sin\left(k_g(z+h)\right)}{\cos\left(k_g h\right)}, & -h < z < 0 \end{cases} \tag{1.4}$$

$$E_z = \frac{\partial_y H_x}{ik_0 \varepsilon_z} = \frac{Ak e^{iky}}{k_0} \begin{cases} \dfrac{1}{\varepsilon_d} e^{-\kappa z}, & z > 0 \\ \dfrac{\cos\left(k_g(z+h)\right)}{\varepsilon_z \cos\left(k_g h\right)}, & -h < z < 0. \end{cases} \tag{1.5}$$

Here k is the spoof plasmon wave vector, k_g is defined by (x), $\kappa = \sqrt{k^2 - \varepsilon_d k_0^2}$, and A is the amplitude. 0, and A are the amplitude. Note that the region $-h < z < 0$ is occupied by the effective medium; thus, ε_y and ε_z are given by (1.1). The area $z < -h$ is chosen to be a perfect conductor, so the fields there are zero. This is a good approximation since $a \ll h$, and losses at the bottom of the grooves are negligible.

By expanding the spoof plasmon wave vector presented in [66], i.e., rearranging it in terms of dielectric parameters of the effective medium, one can obtain the following expression:

$$k = \sqrt{\varepsilon_d k_0^2 + \left(\frac{\varepsilon_d}{\varepsilon_y}\right)^2 k_g^2 \tan^2\left(k_g h\right)}. \tag{1.6}$$

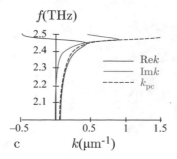

Figure 1.2: The dispersion relation of the spoof plasmons (frequency f vs. wave vector k) supported by a 1D array of grooves with geometrical parameters $a/d = 0.3, a = 3\ \mu m, h = 30\ \mu m$. The real part, $\mathrm{Re}k$, is denoted by the red line and $\mathrm{Im}k$ by the blue line, while the black dashed line is the dispersion relation for a perfect conductor. The dielectric functions ε_d and ε_g are chosen to be one [24].

Here, $k_0 = 2\pi f/c$ is the vacuum wave vector, $h-$ is the grove depth in the semiconductor, k_g is the wave vector of the wave propagating on the grooves defined as [67]:

$$k_g = k_0\sqrt{\varepsilon_g}\left(1 + \frac{l_s(i + 1)}{a}\right)^{1/2}, \tag{1.7}$$

where $l_s = \left(k_0\mathrm{Re}\sqrt{-\varepsilon_s}\right)^{-1}$ is the skin depth.

After substituting Eqs. (1.2) and (1.7) into Eq. (1.6) one obtains the following expression of the wave vector:

$$k = \sqrt{\begin{array}{l}\varepsilon_d k_0^2 + \varepsilon_d^2\varepsilon_g k_0^2 \tan^2\left(\sqrt{\varepsilon_g}hk_0\sqrt{(1 + i)/ak_0\ \mathrm{Re}\left(\sqrt{-\varepsilon_s}\right) + 1}\right)\\ \times\left((1 + i)/ak_0\ \mathrm{Re}\left(\sqrt{-\varepsilon_s}\right) + 1\right)\cdot\left(a/\varepsilon_g - (a - d)/\varepsilon_s\right)^2/d^2.\end{array}} \tag{1.8}$$

Equation (1.8) is the dispersion relation for spoof plasmons supported by the corrugated surface of a semiconductor.

1.2   PROPAGATION OF PLASMONS IN REAL METALS

To illustrate the dependence of the spoof plasmon properties on the metal skin depth, we plot the real and imaginary parts of the wave vector k as the function of frequency f in Fig. 1.2.

The dashed black line in Fig. 1.2 corresponds to the dispersion of spoof plasmons on the perfect conductor. It can be seen that at resonant frequency, which for given parameters is $f \approx 2.5$ THz, the wave vector k_{pc} is infinitely large. However, if losses are taken into account, $\mathrm{Re}k$ (red curve) experiences the backbending similarly to SPPs and has a finite maximum value.

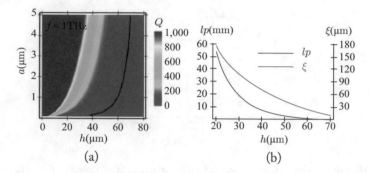

Figure 1.3: (a) Spoof plasmon figure of merit Q as a function of a and h for $f = 1$ THz $(a/d = 0.3)$. Black curve corresponds to $Q = 4\pi$. (b) The propagation length l_p (red curve) and confinement length ξ (blue curve) of spoof plasmons as functions of the depth of the grooves h at $f = 1$ THz for the parameters $a/d = 0.3, a = 3$ μm. Note the different scales on the right and left axes [24].

Also, there is a peak of Imk (blue curve) near the resonance. The analytical approximation for Imk can be obtained from analysis of (1.6)

$$\text{Im}k \simeq \frac{k_0 \varepsilon_d^{3/2}}{2\varepsilon_g} \zeta \left(\varepsilon_g^{1/2} k_0 h \right) \left(\frac{a}{d} \right)^2 \frac{l_s}{a}, \tag{1.9}$$

where $\zeta(x)$ is a function defined as $\zeta(x) = \tan(x) \times \left(s \sec^2(x) + \tan(x) \right)$. This expression is an excellent approximation at all frequencies, except for the region of backbending. As one can see from (1.6), the spoof plasmon propagation length $l_p = 1/(2\,\text{Im}k)$ is inversely proportional to the skin depth of the metal and becomes infinite for the perfect conductor. The propagation length is proportional to d^2 in agreement with numerical computation of Ref. [69].

The spoof plasmon figure of merit $Q = \text{Re}k/\text{Im}k$ is depicted in Fig. 1.3 as a function of the geometrical parameters of the system, a and h at $f = 1$ THz. Physically, Q shows how many oscillations a spoof plasmon undergoes before it decays. The black curve in this figure shows when $Q = 4\pi$, which corresponds to energy decay length equal to wavelength. The area to the left of this curve is suitable for propagation of THz field with spoof plasmons.

Besides the figure of merit, Q, the other important propagation characteristics of spoof plasmons are the energy attenuation length (propagation length) $l_p = 1/(2\,\text{Im}k))$ and the energy confinement in the dielectric $\xi = 1/(2\,\text{Re}\kappa)$. In Fig. 1.3, these two parameters are shown as a function of the depth of the grooves, h. Note that the better the confinement is, the lower the propagation length.

CHAPTER 2

Examples

2.1 SI

In this chapter, a simple example of polaritons in corrugated semiconductors at THz frequencies is given. To illustrate the properties of SPPs we plot the wave vector k (Eq. (1.5)) as a function of the frequency. The case of a heavy-doped Si is considered, assuming that the doping level is $N = 5 \cdot 10^{19}$ cm^{-3} [70]. It is assumed that the structure is surrounded by air, i.e., $\varepsilon_g = \varepsilon_d = 1$.

The dispersion curves of spoof SPPs on structured surfaces with $d = 10$ μm are shown in Fig. 2.1a. It is of particular interest to analyze the effect of the groove depth on the dispersion curves of spoof SPPs. For this reason two different groove depths, i.e., $h = 3d$ and $h = 2.5d$, are suggested for the study. It should be pointed out that several values of the groove width are considered for each groove depth. As seen from Fig. 2.1, the asymptotic frequency decreases by increasing the groove depth. This is also the case for the groove width. However, the groove depth drastically affects asymptotic frequency compared to the groove width. The losses of these spoof SPPs as a function of frequency are presented in Fig. 2.1b. It can be observed that the loss of spoof SPPs grows significantly with increasing the frequency. For instance, in the case of $a = 0.4d$ and $h = 2.5d$, the attenuation coefficient of the spoof SPPs is 23 m^{-1} for $f = 0.6$ THz, and it increases up to 59 m^{-1} for $f = 0.8$ THz and 134 for $f = 1$ THz, respectively. Compared to a metal THz waveguide described in [69], the corrugated semiconductor surface exhibits relatively high loss for guiding THz wave. The mentioned outcome appears due to the fact, that the permittivity of metal is smaller that that of silicon. The losses in our study are assigned by investigating a skin effect, i.e., introducing the skin depth, i.e., a measure of how deep electromagnetic wave can penetrate into a material. It is well known that the penetration depth of a semiconductor material, influencing the losses [66], depends on permittivity [21].

The analysis of the effect of the lattice constant (d) on the dispersion of spoof SPPs is of particular importance. Figure 2.2a shows the dispersion curves for spoof SPPs on corrugated surfaces with different lattice constants $d = 5, 7$, and 10 μm, respectively. The groove parameters are $a = 2$ μm and $h = 30$ μm for all cases. The losses of spoof SPPs for three cases are plotted in Fig. 2.2b. As seen from Fig. 2.2b, a smaller lattice constant corresponds to a larger loss of spoof SPPs for a given frequency. As shown in Fig. 2.2b, the lattice constant drastically affects losses of the spoof SPPs. For a given frequency, an increase of the lattice constant may result in a significant reduction of the loss of spoof SPPs. With the need for a compact, reliable, and flexible THz system for various applications, a low-loss THz waveguiding system is essential.

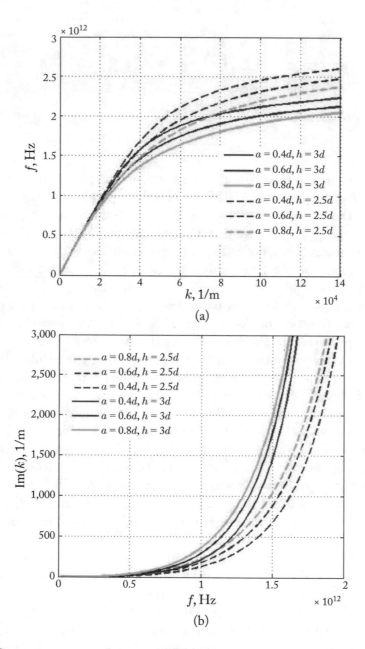

Figure 2.1: (a) Dispersion curves for spoof SPPs. (b) Attenuation coefficients of spoof SPPs. lattice constant $d = 10\ \mu$m.

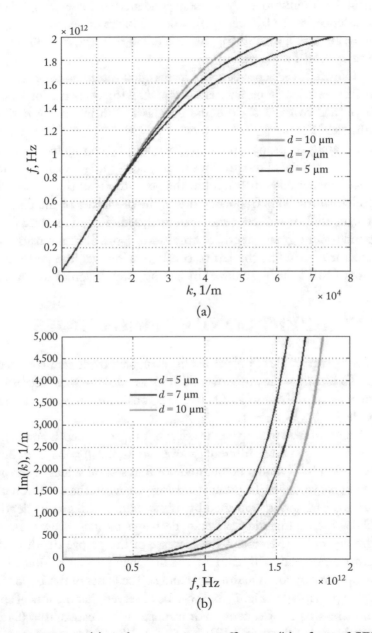

Figure 2.2: Dispersion curves (a) and attenuation coefficients (b) of spoof SPPs for different lattice constants $d = 5, 7$, and $10\ \mu$m, respectively. Parameters of grooves $a = 2\ \mu$m and $h = 30\ \mu$m.

It is of particular interest to analyze the dependences of losses vs. guide parameters. The investigated dependences at the frequency $f = 0.5$ THz are presented in Fig. 2.3. The groove parameters are $d = 10$ μm, $h = 2.5d$ for the case (a), $a = 2$ μm, $h = 30$ μm for the case (b), and $d = 10$ μm, $a = 0.4d$ for the case (c).

As shown in Fig. 2.3, the dependence of losses upon the guide parameter a is linear, while it is exponential-like with guide parameters d and h. Moreover, the losses can be decreased by increasing the lattice constant d. The loss decreases with increasing d due to the weaker confinement of the field [21].

A method for analyzing spoof SPPs on a periodically corrugated semiconductor surface has been presented. Compared to the previous works, our approach enables us to investigate the semiconductor case, since it takes into account the permittivity of the semiconductor expressed by the Drude model. In the THz frequency range, the properties of the dispersion and loss of spoof SPPs on corrugated Si surfaces have been analyzed. The asymptotic frequency of spoof SPPs mainly depends on the groove depth, but the loss of spoof SPPs is sensitive to all parameters of the surface structure, including the lattice constant. However, the performance of low-loss propagation for spoof SPPs can be achieved by an optimum design of the surface structure.

2.2   TRANSPARENT CONDUCTING OXIDES

In this section, a simple example of polaritons in corrugated oxide semiconductors at THz frequencies is given. To illustrate the properties of SPPs we plot the wave vector k (Eq. (1.5)) as a function of the frequency. The case of ZnO (AZO) is considered. It is assumed that the structure is surrounded by air, i.e., $\varepsilon_g = \varepsilon_d = 1$.

It is of particular interest to analyze the effect of the groove depth on the dispersion curves of spoof SPPs. For this reason three different groove depths, i.e., $h = 3d, h = 2.5d$, and $h = 2d$, are suggested for the study. It should be pointed out that several values of the groove width are considered for each groove depth. In other words, we analyzed the evolution of the dispersion relation of the surface EM waves supported by this structure with groove depth and width. As seen from Fig. 2.4a, the asymptotic frequency decreases by increasing the groove depth and decreasing the groove width. Moreover, the increase of the groove depth leads to the modes occupying the narrow bandwidth. To sum up, the dispersion curves exhibit SPP-like behavior, presenting an asymptote controlled mainly by a and h. The losses of these spoof SPPs as a function of frequency are presented in Fig. 2.4b. It can be observed that the loss of spoof SPPs grows significantly with increasing the frequency. For instance, in the case of $a = 0.4d$ and $h = 2.5d$, the attenuation coefficient of the spoof SPPs is 5.88×10^{-7} m^{-1} for $\omega = 1.2$ eV, and it increases up to 2.56×10^{-6} m^{-1} for $\omega = 1.8$ eV and 3.37×10^{-6} for $\omega = 2$ eV, respectively. Compared to a semiconductor structure described in [77], the corrugated transparent conducting oxide surface exhibits relatively low loss for guiding the wave. The mentioned outcome appears due to the fact that the permittivity of the transparent conducting oxide is smaller than that of silicon. The

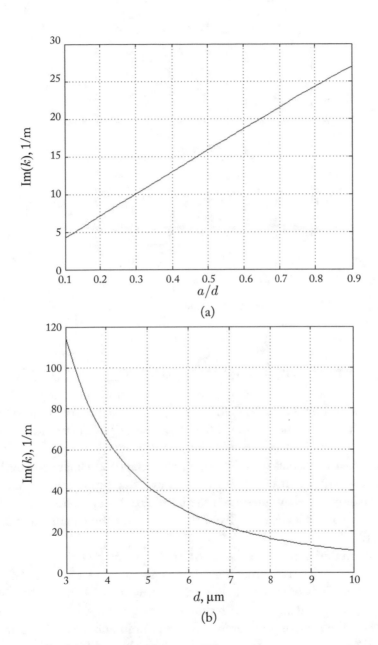

(a)

(b)

Figure 2.3: Losses vs. guide parameters. (*Continues.*)

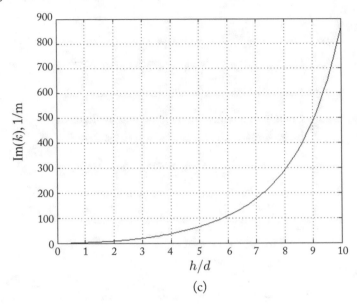

(c)

Figure 2.3: (*Continued.*) Losses vs. guide parameters.

losses in our study are assigned by investigating a skin effect, i.e., introducing the skin depth, i.e., a measure of how deep electromagnetic wave can penetrate into a material.

It should be mentioned that the field is strongly confined to the surface. This is displayed in Fig. 2.5, where the distribution of the E field in a unit cell of the surface structure is plotted for the case of $a = 0.4 \cdot d$, $h = 3 \cdot d$, $d = 1$ nm.

In addition to the dispersion relation, k, the energy attenuation length (propagation length) denoted as $l_p = 1/(2 \, \mathrm{Im}k)$ [66] is the other essential propagation characteristic dealing with the spoof plasmon. We have presented the normalized propagation length as the function of the depth of the grooves in Fig. 2.6. Note that the better the confinement is, the lower the propagation length.

To study the material effect on the spoof SPP wave propagation, we calculated the dispersion characteristics for different materials. The analysis of the effect of the material on the dispersion of spoof SPPs is of particular importance. Figure 2.7a shows the dispersion curves for spoof SPPs on corrugated surfaces with different materials, i.e., AZO and indium tin oxide (ITO), respectively. The groove parameters are $a = 0.4$ nm and $h = 1.2$ μm for all cases. The losses of spoof SPPs for two cases are plotted in Fig. 2.7b. As seen from Fig. 2.7b, a material with a higher permittivity, i.e., AZO corresponds to a larger loss of spoof SPPs for a given frequency. As shown in Fig. 2.7b, the background permittivity drastically affects losses of the spoof SPPs. For a given frequency, a decrease of the lattice constant may result in a significant reduction

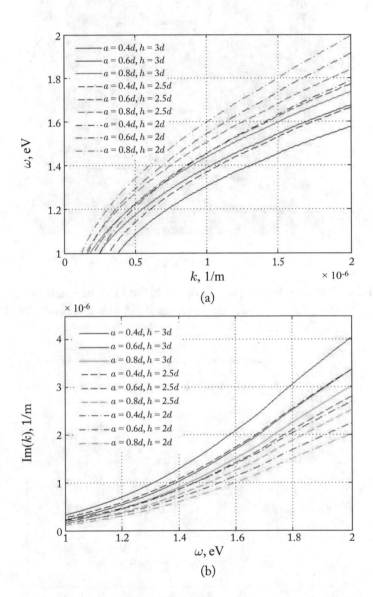

Figure 2.4: (a) Dispersion curves for spoof SPPs. (b) Attenuation coefficients of spoof SPPs. Lattice constant $d = 1$ nm.

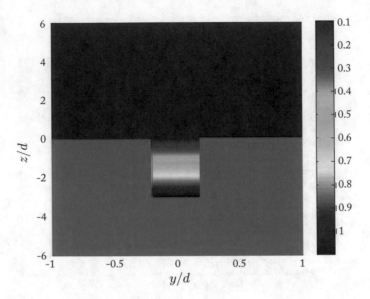

Figure 2.5: Spatial variation of the amplitude of the E field in a unit cell of the surface structure with $d = 1$ nm, $a = 0.4 \cdot d$, $h = 3 \cdot d$. The colorbar illustrates the normalized values of the E field.

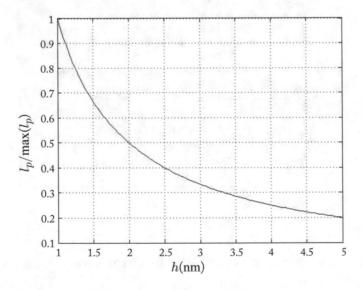

Figure 2.6: The normalized propagation length of spoof plasmons as a function of the depth of the grooves h at $w = 1.4$ eV for the parameters $d = 1$ nm, $a = 0.4d$.

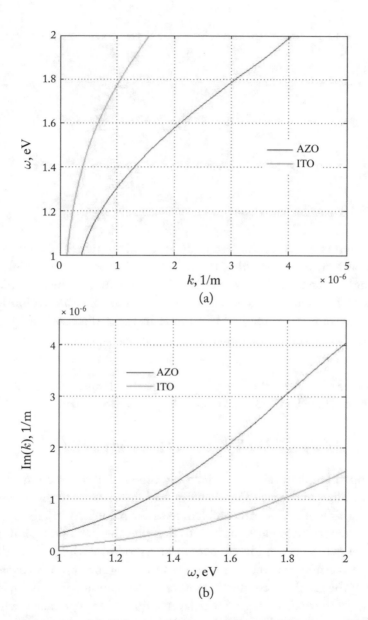

Figure 2.7: Dispersion curves (a) and attenuation coefficients (b) of spoof SPPs for different materials, i.e., AZO and ITO, respectively. Parameters of grooves $a = 0.4$ nm and $h = 1.2$ nm.

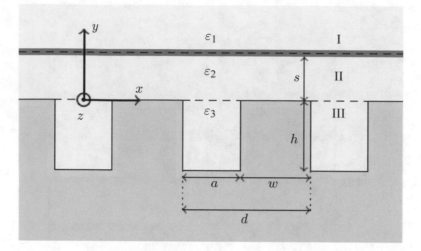

Figure 2.8: The schematic view of the system under consideration. The grey area depicts a substract where the perfect metal thin film, in gold color, is deposited. The dielectric regions are depicted by the light-blue areas. Graphene sheet is represented by the dark-blue thin area. All the different geometrical parameters and dielectric functions in either region I–III are marked in the figure. The system is assumed to be infinite in the z-direction and periodic in the x-direction [132].

of the loss of spoof SPPs. With the need for a compact, reliable, and flexible THz system for various applications, a low-loss THz waveguiding system is essential.

2.3   METAL COATED WITH GRAPHENE

The usage of a doped graphene sheet to effectively tune the plasmonic properties of the spoof plasmons, by controlling graphene's Fermi energy and the distance to the grooved metallic surface is proposed in this work. First, this problem has been solved in [78], for a 2D metal grating, by means of the effective medium approach to describe the periodic region of the structure. Herein, a 1D metal grating is considered. Moreover, a mode-matching method considering the metal to be perfect is employed. The former proves to be a good approximation. Moreover, the non-local effects in the graphene conductivity have been accounted for employing the Mermin's formula. The usage of the additional graphene layer seems to be very efficient when the energy of the uncoated diffraction grating SSPs is close to the energy of the graphene plasmons, especially in the THz spectral range. The mentioned tunability properties can be applied for waveguiding and sensing applications.

A semi-infinite grooved perfect-conducting surface coated by a graphene sheet parallel to its surface, like the one depicted in Fig. 2.8 is considered.

The dielectric area is divided in three different regions (I, II, and III) with different permittivities ($''1$, $''2$, and $''3$, respectively). It is assumed that the grooves in the metal are rectangular, with width a, depth h, and period d. It is assumed that the metal itself is perfect. The graphene sheet is located at a distance s from the top of the grooves, and is characterized by a non-local conductivity σ dramatically affected by its Fermi energy E_F and relaxation energy Γ.

2.3.1   COMPUTATION OF PERMITTIVITY AS THE FUNCTION OF FREQUENCY AND BIAS FOR ANISOTROPIC COMPOSITES FOR NON-GRAPHENE CASE

Aiming to estimate the limits of attenuation control at a given operating frequency, one needs to calculate the bias-affected reflection and transmission coefficients of a plane screen of optimal for this frequency thickness and composition. It is impossible to randomly choose the operating frequency, as it is located in the vicinity of FMR frequency of the available permeable wire. Below the properties of a screen filled with Fe-based wires are estimated, as their magnetocapacitance effect (Fig. 2.11) exceeds that of Co-based ones (Fig. 2.10).

The reflection R and transmission T coefficients of a plane sample are dramatically affected its complex permittivity ε (in case under study $\mu \approx 1$) and thickness [79]. Considering the permittivity of a diluted composite filled with insulated metal inclusions we generalize the approach [80] to microwave absorption of a matrix mixture upon the mixture with elongated magnetoresistive ellipsoids. One can calculate the complex permittivity of an anisotropic composite filled with aligned wire sections by means of Maxwell–Garnett mixing formula (2.1):

$$\varepsilon_{mix} = \varepsilon_h \left[1 + \frac{(\varepsilon_{\text{sec}} - \varepsilon_h)\, p}{(1 - p)N\, (\varepsilon_{\text{sec}} - \varepsilon_h) + \varepsilon_h} \right]. \tag{2.1}$$

Here $\varepsilon_h \approx 1.2$ is the binder permittivity; N is the depolarization factor of the wire of length L and diameter d; and p is the volume fraction of wire sections of permittivity ε_{sec}. The latter is to be determined as follows.

The wire is treated as a complex dielectric with effective permittivity ε_{eff}. In the case under consideration, the section length L is not negligible compared to the operating wavelength λ. Consequently, it is impossible to implicitly characterize the wire inclusion as a small ellipsoid by the shape-dependent factor N in Eq. (2.1). The relatively long ($L/\lambda > 0.1$) wire is a dipole. Calculating the complex permittivity ε_{sec} of a dipole, the serial connection of all constituents (capacitive, inductive and resistive) of dipole admittance is considered. The first item in denominator of relation (2.2) corresponds to capacitive resistance of a dipole. The second one corresponds to its inductance (see [80, 81]), while the third item corresponds to pure resistance of the wire. The radiation loss $R_{rad} = 75$ Ohm [80] should be added to the third item, if the aligned dipoles are characterized by the random distribution. Therefore, the effective permittivity of the wire section ε_{sec} is drastically affected by the properties of the wire as well as by the section length and distribution of sections in matrix:

$$\varepsilon_{sec} = \frac{\lambda^2}{\frac{\lambda^2}{\varepsilon'_{eff}} - \pi d^2 \ln\left(\frac{L}{d}\right) + i\left(\frac{\lambda^2}{\varepsilon''_{eff}} + \frac{\pi^2 c \varepsilon_{vac} \lambda d^2}{2L} R_{rad}\right)}, \qquad (2.2)$$

where λ is the wavelength; $c = 3 \times 10^8$ m/sec is the light velocity; $\varepsilon_{vac} = 8:85 \times 10^{-12}$ F/m is the dielectric constant; and ε_{eff} is the effective permittivity of the infinitely long wire.

Note that ε_{eff} of an inclusion of complex dielectric is influenced by both frequency and inclusion size even in case the latter being much smaller than the wavelength. In case of the inclusions being electrically thick [80] (the smallest dimension of inclusion is comparable or higher than the penetration depth δ), the effect of inclusion size upon the microwave permittivity of a composite becomes significant.

If the metal wire of d.c. conductivity σ is very thin ($d \ll \delta$), the imaginary part of its permittivity ε'' is proportional to the wavelength λ while the real part ε' is about unity:

$$\varepsilon = 1 - i\sigma\lambda/2\pi\varepsilon_{vac}c. \qquad (2.3)$$

If the wire is thick enough ($d \geq \delta$), the fact that the effective conductivity σ is frequency-dependent as it is the function of penetration depth has to be taken into account. It is well known that the penetration depth of a magnetic material is influenced by both complex permittivity $\varepsilon = \varepsilon' - i\varepsilon''$ and permeability $\mu = \mu' - i\mu''$. Consequently, the effective permittivity ε_{eff} of a long permeable wire parallel to the microwave electric field is calculated considering the penetration depth in the infinite cylinder with both magnetic and dielectric losses [82, 83]:

$$\varepsilon_{eff} = \frac{\varepsilon J_1\left(2\pi d\sqrt{\varepsilon\mu}/\lambda\right)}{J_0\left(2\pi d\sqrt{\varepsilon\mu}/\lambda\right) 2\pi d\sqrt{\varepsilon\mu}/\lambda - J_1\left(2\pi d\sqrt{\varepsilon\mu}/\lambda\right)}. \qquad (2.4)$$

Here, J_0 and J_1 are Bessel functions; μ is the circumferential permeability of metal wire; permittivity ε is defined by (1.3). It is worthwhile noting that in the vicinity of FMR the frequency dependence of complex permittivity ε_{eff} for a wire of permeable metal differs significantly from that of impermeable one.

Aiming to obtain the FMR parameters and $\mu(f)$ the simplest model of a single-domain inclusion with uniaxial anisotropy with the bias and anisotropy fields being parallel insidered. In this case, the FMR spectrum possesses the Lorenzian shape characterized by Eq. (1.5) similar to derived in [84]:

$$\mu = 1 + \frac{(\mu_{st} - 1)/(1 + H_{ext}/H_A)}{1 - \left[\frac{2\pi f}{\gamma(H_A + H_{ext})}\right]^2 + i\frac{2\pi f \Gamma}{\gamma(H_A + H_{ext})}}, \qquad (2.5)$$

where $\gamma = 2.8$ GHz/kOe is the gyromagnetic factor for Fe; μ_{st} is the quasistatic (low-frequency) permeability under zero bias ($H_{ext} = 0$); H_a and H_{ext} are the anisotropy and the external bias fields correspondingly; Γ is the damping factor; and f is the frequency.

The quasi-bulk permeability of metal defining the penetration depth together with its bulk conductivity, inclusion shape, and orientation (1.4) is characterized by Eq. (1.5). Equation (1.5) is easy to comprehend taking into account that the longitudinal demagnetizing factor is close to zero as $L \gg d$ and that the induced current magnetizes the wire circumferentially with demagnetizing factor also equal to zero. The approach is justified at least for the Fe-based wires with positive magnetostriction where the domains are magnetized mostly axially. The fraction of domains magnetized perpendicularly to the axis is small [85].

According to the data in Fig. 2.9b the FMR frequency of the wire is about $F_{rez} \approx 9.2$ GHz. However, for quantitative estimations one needs all parameters of FMR spectrum. We reconstruct the FMR spectrum treating the data [86] on the conductivity of a similar Fe-based wire as a function of frequency.

Figure 2.13 shows the spectrum of circumferential permeability $\mu(f)$. The former is reconstructed by fitting parameters of Eq. (1.5) to minimize the discrepancy of the effective conductivity $\sigma_{eff} = 2\pi f \varepsilon_{vac} \varepsilon''_{eff}$ calculated using Eq. (1.4) from that of the measurements [86]. The fitted parameters of permeability spectrum (Fig. 2.13) are as follows: $\Gamma = 0:04$; $\mu_{st} = 2:5$; $H_a = 3:25$ kOe.

The zero-bias FMR frequency F_{rez} for these parameters is equal to 9 GHz, which is close to 9.2 GHz estimated from data in Fig. 2.11. The product of static permeability and resonance frequency $(\mu_{st} - 1) \times F_{rez} \approx 13:5$ GHz calculated using Eq. (1.5) is lower than Snoek's limit, which is about 40 GHz for pure iron. The difference may be attributed to dilution of iron, as the FeSiBMnC amorphous alloy is a solid solution. The value of damping factor is typical for glass-coated wires [87, 88]. The penetration depth δ in Fig. 2.13 is obtained based on the reconstructed permeability (1.5) and permittivity of the infinite wire (1.3).

If the wire permeability is frequency independent (the well known particular case of an impermeable alloy), the penetration depth decreases with frequency as $\delta(f) \sim 1/\sqrt{f}$. In case of a permeable metal, the function $\delta(f)$ is more complicated and is dramatically affected by the shape of FMR spectrum.

One can easily figure out that there are two extremums of $\delta(f)$ function in case of the Lorenzian shape of the line of magnetic absorption (1.5). Namely, the skin-depth δ is minimal $\delta_{min} = 1/\sqrt{2\pi F_{rez} \sigma \mu_{vac} (\mu_{st} - 1)}$ at the FMR frequency F_{rez}, and δ is maximal $\delta_{max} = 1/\sqrt{\sigma \mu_{vac} \pi f (|\mu| + \mu'')}$ at the frequency where $(|\mu| + \mu'')$ is minimal. Here $\mu_{vac} = 1.25 \times 10^{-6}$ Hn/m is the magnetic constant. The highest relative resistance of the wire corresponds to the minimal skin-depth δ and vice versa.

It is worthwhile noting that these very extremums of penetration depth specify the maximal change of wire impedance and consequently define the optimal parameters of a tunable screen, namely the composition, operating frequency, necessary bias strength, and obtained range of attenuation control. The frequency of δ_{min} belongs to the limits of experimental frequency range and is determined relatively accurately. Doing so, a good agreement between the experimental and simulated data at 10 GHz (Figs. 2.11 and 2.12) is obtained. The frequency

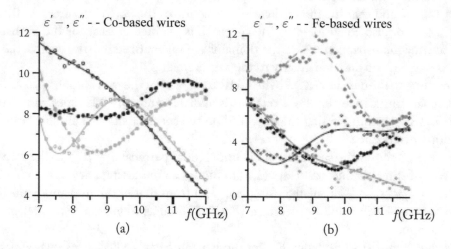

Figure 2.9: Permittivity spectra of samples filled with 10 mm-long: (a) Co-based wires; (b) Fe-based wires. The black lines correspond to zero bias, the gray ones correspond to 1 kOe bias applied parallel to the microwave electric field.

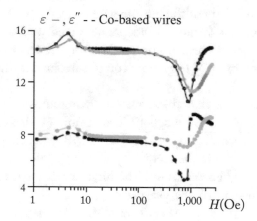

Figure 2.10: Permittivity as bias function for the sample with Co wires ($f = 6$ GHz, $L = 10$ mm).

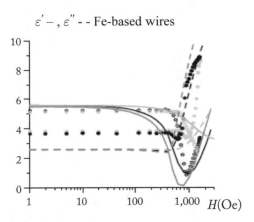

Figure 2.11: Permittivity as function of bias for the sample with Fe-based wires ($f = 10.4$ GHz, $L = 10$ mm). The experimental data are presented by dots, the simulation data (see Section 2.5) are presented by continuous lines. The black dots and lines correspond to bias applied parallel to the microwave electric field $H_{ext} \parallel E$; the gray dots and lines correspond to perpendicular bias $H_{ext} \perp E$. The light-gray and black lines correspond to randomly oriented wires; the dark-gray lines correspond to wires aligned parallel to E.

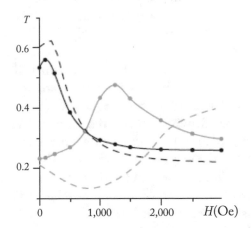

Figure 2.12: Power transmission as function of parallel bias for the sample with Fe-based wires ($L = 10$ mm, $f = 10$ GHz, and $f = 15$ GHz, black and gray lines correspondingly). The dots connected by solid lines present the experimental data, while the dashed curves present the results of simulation (see Section 2.5).

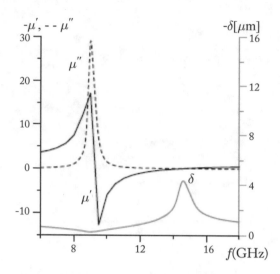

Figure 2.13: Reconstructed frequency dependence of permeability (solid and dashed black lines) and skin-depth (solid gray line) for FeSiBMnC microwire with $\mu_{st} = 2 : 5$, $\Gamma = 0 : 04$, $\sigma = 70000$ Ohm^{-1}cm^{-1}, and $Ha = 3 : 25$ kOe.

of δ_{max} is estimated taking into account the Lorenzian shape of FMR spectrum. It is demonstrated that the estimated value exceeds the actual frequency of δ_{max} by the comparison of the experimental and simulated transmission data at 15 GHz (Fig. 2.12).

Note that under $H_{ext} = 0$ the wire resistance $\delta_{eff} = 1/\sigma_{eff} = 1/2\pi f \varepsilon_{vac}\varepsilon''_{eff}$ is minimal at 14.4 GHz, while under $H_{ext} = 1.8$ kOe the resistance is maximal. The magnetoimpedance curve at 14.4 GHz is shown further in Fig. 2.16. The impedance change here requires high bias, but the change is about four times higher than that due to GMI [83].

Aiming to obtain the permittivity close to the measured values ($\varepsilon' \approx 12$), the volume fraction p in Eq. (1.1) is expected to be twice smaller than the filling factor for the manufactured plane-isotropic samples, where the wires are not aligned. The calculation produces the value $p \approx 0 : 0026\%$ that is about 3 times lower than the filling factor of the samples under study. The discrepancy can be attributed to the wire sections nonparallel to the plane of the sheet and to distortion of wire properties at the section ends [89].

Finally, one can calculate the permittivity spectra of a composite under bias of given strength (Fig. 2.14) based on the spectrum of circumferential permeability for Fe-based wire (Fig. 2.11), wire conductivity (70,000 Ohm^{-1}cm^{-1}), length ($L \approx 1$ cm), and core diameter ($d = 4$ μm). These spectra are the result of the interference between the wire-dipole resonance and the current-induced resonance of circumferential permeability (magnetoimpedance resonance). The transformation of the effective permittivity spectra from a resonance type to a relaxation one due to GMI in the vicinity of the antenna resonance [90] is clearly demonstrated

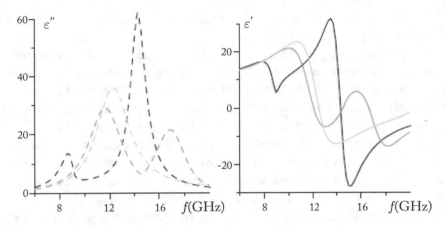

Figure 2.14: Permittivity ε' and ε'' (continuous and dashed lines) dispersion curves under bias $H = 0, 2, 4$ kOe (black, dark-gray, and light-gray curves correspondingly) of a composite containing 2.7×10^{-3} vol% Fe-based wire sections 8 mm long.

by the spectra in Fig. 2.14. But in the case under consideration, the interference of antenna resonance and impedance resonance is optimized. Consequently, the Debye absorption corresponding to high-impedance state of the wire is transformed into two lines, and the total permittivity change is significantly higher than that due to GMI effect [90].

The highest change of the wire resistance is observed at the frequency about 14.5 GHz, where the skin depth reaches its maximum under zero bias (see Figs. 2.13 and 2.16). At this frequency, the wire has minimal effective resistance, and the absorption line has the minimal width (black lines in Fig. 2.14). The second absorption peak at about 8.5 GHz has the relaxation nature (the real part of wire resistance is much higher than the inductive one), as at this frequency the wire resistance is maximal due to zero-bias FMR. All together, the deviation from FMR frequency abruptly decreases the wire resistance and makes the absorption line much sharper than the relaxation (Debye) line.

The wire resistance reaches maximum under 2 kOe bias. The former shifts the FMR spectrum to 14.4 GHz (see Figs. 2.13 and 2.16). Under this bias the absorption exhibits the relaxation nature similar to absorption in an RC circuit. The losses are minimal because the wire resistance is too high (dark-gray lines in Fig. 2.14). The frequency deviation from the peak of FMR spectrum results in the decrease of effective resistance. Moreover, the two separate lines of dielectric absorption are formed. The further increase of bias strength again decreases the wire resistance at 14.4 GHz and increases the ε'' value. Under 4 kOe bias, the wire permeability within the investigated frequency band is close to unity. However the effective resistance is higher than in the case of maximal skin-depth. Therefore, the absorption peak has lower frequency (about 12.5

GHz), and the absorption line is much wider than in the case of minimal resistance (compare the black and light-gray curves in Fig. 2.14).

The two-line shape of permittivity spectra in Fig. 2.14 resembles the spectra for a FeCoSiB microwire stretched across a coaxial line [91], although the magnetocapacitance effect is much stronger in the case under study. The reason is that in [91] the frequency of dielectric resonance ($\sim$ 1 GHz) is lower than the FMR frequency (there are no data on the wire permeability in [91], but we suppose from the permittivity spectra that the peak of magnetic loss takes place at $\sim$ 2 GHz). Therefore, the step of wire impedance takes place at a shoulder of absorption curve, where its effect on the permittivity is small. In case of the samples under study, the frequency of dielectric absorption is tuned to be equal to the frequency of minimal wire impedance. Therefore, the magnetoimpedance effect on the permittivity is resonantly increased.

The composition and structure of the optimized screen differs from that of the samples studied experimentally. Aiming to obtain the widest range of attenuation control one has to engineer the length of the wire section to the operating frequency (14.5 GHz in Fig. 2.14) and to align the sections parallel to electric and bias fields. Application of the simplified model that assumes the Lorenzian FMR spectrum and considers the idealized structure with parallel wires results in the discrepancy between the calculated and experimental dispersion curves (Figs. 2.9 and 2.14).

Aiming to take into account the random orientation of wire sections in the measured sample one needs to consider the effective magnetizing field as the function of the angle α between the wire axis and bias field: $H_{eff} = H_A + H_{ext} \cos \alpha$. For the sake of simplifying calculations the contribution of domains that are magnetized perpendicular to the wire axis is neglected. It is assumed that the wire remains magnetized axially independently on bias orientation due to the demagnetizing factor perpendicular to the wire axis being equal to zero. Consequently, aiming to describe the circumferential permeability as the function of the angle α between the wire axis and the bias field one has to substitute H_{ext} in Eq. (1.5) by $H_{ext} \cos \alpha$ in case of a perpendicular bias H_{ext} is replaced by $H_{ext} \sin \alpha$.

It is worthwhile noting that under zero bias the conductivity of wires is independent of their orientation. However, under non-zero bias the circumferential permeability and consequently the wire conductivity is the function of the angle α between the wire axis and bias. Hence, the electric polarization of the wires as the function of the same angle α between the wire axis and the microwave electric field is considered. Therefore, we modify Eq. (1.1):

$$\varepsilon_{mix} = \varepsilon_h \left[1 + \int_0^{\pi/2} \frac{[\varepsilon_{wire}(\alpha) - \varepsilon_h] \, p \cos^2 \alpha}{(1 - p \cos^2 \alpha) \, N \, [\varepsilon_{wire}(\alpha) - \varepsilon_h] + \varepsilon_h} d\alpha \right]. \tag{2.6}$$

Using (1.1), (1.6), and the modified Eq. 1.5, the curves that resemble the experimental dependence of permittivity on bias (see the black curves and dots in Fig. 2.11) are observed. The discrepancy between the calculated permittivity for plane-isotropic sample and experimental data (the experimental permittivity changes steeper with bias increase, ε' reaches minimum at

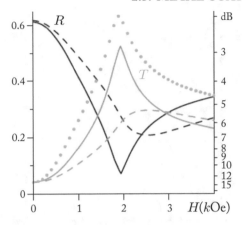

Figure 2.15: Power reflection (black lines) and transmission (gray lines) coefficients at 14.5 GHz as bias function of 0.4-mm-thick screen filled with wire sections 8 mm long. Solid lines correspond to the screen with aligned wires; dashed ones correspond to random oriented wires. The gray dots represent transmission in case of regular distribution of aligned wire sections in 0.25-mm-thick screen.

bias 300 Oe higher) may arise because of following reasons. The actual shape of FMR spectrum may differ from the Lorenzian one. The width of FMR spectrum (the effective damping factor Γ) is usually bias-dependent [92]. The domain structure at the ends of the wire sections is known to be distorted [85, 93]. Consequently, the FMR parameters in the middle and at the ends of wire sections are different. The conductivity and domain structure across the section of the wires may be non-uniform, etc.

Using the calculated dispersion curves (Fig. 2.14) and applying the Fresnel's relations one can calculate the reflection and transmission spectra [79] of a sheet filled with parallel sections of fibers under study, which has complex permittivity ε_{mix} (Fig. 2.15). At the frequency corresponding to the resonance of wire length (14.4 GHz for 9-mm-long sections) $\tan \delta_\varepsilon = \varepsilon''/\varepsilon' \approx 1$. Thus, the transmission coefficient gradually decreases with increase of sample thickness (the interference contribution is negligible). The transmission loss for a transparent screen is defined by the product of screen thickness and filling factor. If the admissible loss is equal to 3 dB, then the range of attenuation control for a sample with the wires under study is about 10 dB (Fig. 2.15).

The regular distribution of aligned wire sections ($R_{rad} = 0$) for the same attenuation of the opaque screen decreases the transmission loss in the transparent state for about 1 dB compared to a disordered anisotropic composite (see dotted and gray lines in Fig. 2.15).

Note that at the resonance frequency of wire sections ($\sim$ 14.4 GHz for 9-mm wires) the real part of mixture permittivity ε'_{mix} under 0–2 kOe bias is close to unity while the imaginary part is high $\varepsilon''_{mix} > 60$ and depends on the bias strength (see Fig. 2.13). Thus, at the resonance

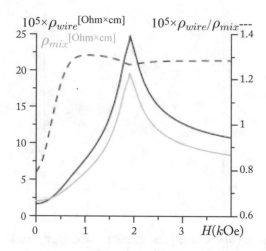

Figure 2.16: Effective resistance of the wire at 14.5 GHz, the equivalent resistance of the composite and their ratio as bias function presented by solid black, solid gray, and dashed black lines correspondingly.

frequency the composite with wire sections behaves like a resistive substance, where its equivalent resistance increases with bias, or like a composite similar to one described in [91], where the wire sections are interconnected into a continuous lattice.

The dependence of transmission on bias (Fig. 2.15) is similar to the dependence of equivalent resistance of a sample ρ_{mix} on bias and the dependence of wire resistance ρ_{wire} on bias (Fig. 2.16): $\delta_{eff} = 1/2\pi f \varepsilon_0 \varepsilon''$, where ε'' of the wire is defined by Eq.(1.4), and ε'' of a sample is defined by (1.1).

Comparing the effective resistance of a single wire ρ_{wire} and the equivalent resistance ρ_{mix} of the sample at 14.4 GHz it is observed that the ratio of wire resistance to equivalent resistance of the sample steeply increases with the bias increase up to ~ 800 Oe. Under higher bias the ratio ρ_{wire}/ρ_{mix} is almost constant. It is easy to comprehend the Γ-shape of curve by comparing it with resistance data (Fig. 2.16) and permittivity dispersion curves in Fig. 2.14. Under low bias the wire sections behave like resonant half-wave dipoles; the damping factor is small; and the magnetoimpedance effect is increased proportionally to the current passing through a wire section. The current through a wire section is amplified compared to the current through a continuous wire due to dipole resonance. Under bias exceeding 1 kOe the resonance is damped, and the sections behave like resistive fibers that can be interconnected into a continuous lattice penetrating the sample (the structure measured in [91]). In this case, the ratio ρ_{wire}/ρ_{mix} is bias independent. Consequently, the effect of bias on attenuation of a sample with identical wire sections is amplified resonantly in comparison with the effect in the sample with continuous wires.

The above estimations of the limits of attenuation control are valid for a sample with aligned wire sections. The anisotropic composite is difficult to manufacture in practice, but the screen with random-oriented wires exhibits much lower range of attenuation control (dashed lines in Fig. 2.15). In this case, the maximal transparency is reached under higher bias. The main problem in this case is that the transmission loss in a transparent state is approximately twice higher than that for a screen with aligned wires. The increased loss arises from the higher conductivity of the less magnetized wires that are inclined to bias field.

The transmission curves at 10 and 15 GHz for the actual plane-isotropic sample with 10 mm-long wires (Fig. 2.12) are calculated applying the similar formalism (Eqs. (2.1), (2.5), (2.6)). A good agreement between the experimental and simulated data at 10 GHz as the FMR frequency (the frequency of δ_{min}) has been obtained. Moreover, the anisotropy field H_A (1.5) is determined relatively accurately. The frequency of δ_{max} and the value of minimal resistance of the wire are estimated assuming the Lorenzian shape of FMR spectrum. It is indicated that the actual FMR spectrum may be of asymmetric shape taking into consideration the discrepancy between the experimental and simulated transmission data at 15 GHz. The above mechanism of permittivity control opens the wide avenues for the largest possible control range as it is based on the switch from the maximal to the minimal effective conductivity of the wire. The width and shape of FMR spectrum dramatically affect the bias strength and permittivity change: the lower the damping factor (Γ in Eq. (2.5), the lower the bias and the higher the transparency control range. For a given width of FMR spectrum, the lower the FMR frequency, the nearer the frequencies of minimal and maximal surface impedance of the wire (Fig. 2.11). Therefore, the lower the operating frequency of a composite screen, the lower the necessary bias as the less is the required shift of FMR spectrum. For the FMR frequency of 9 GHz, the switching bias is about 2 kOe, while for ~ 2 GHz the switching bias is about $500 \div 700$ Oe [91].

One can reach the maximal control range only in case of three conditions being satisfied at the same time. Namely, the operating frequency must be equal to the frequency of maximal zero-bias conductivity; the wire length must be equal to $\lambda/2$ at this very frequency; the maximal penetration depth δ must be close to wire diameter. The higher the permittivity ε_h of the binder in which the wires are immersed, the shorter the physical length of a half-wave dipole, the lower the dipole inductance and the wider the line of dielectric absorption. Therefore, the lower the attenuation of the opaque screen, the lower the range of attenuation control. It is important that if the FMR spectrum is wide or its shape is asymmetrically distorted, the maximum of effective conductivity may be reached at the frequency higher than the estimated one or may be totally absent. The case resembles the isotropic sample: the range of transparency control is smaller and the control needs stronger bias.

2.3.2   DISPERSION RELATION

The procedure adopted to characterize the spoof plasmons consists on the modal decomposition of their fields in a Fourier series. This method has been applied before in [94, 95], however it has not been considered for the graphene case. Let us start with the solution of the wave equation retrieved from Maxwell's equations in either region I, II, or III. Taking into account explicitly p-polarized modes with a frequency ω and a harmonic time-variation $e^{-i\omega t}$, the magnetic field in region λ must have the form $\mathbf{B}_\lambda(\mathbf{r}, t) = B_\lambda(x, y)e^{-i\omega t}\hat{\mathbf{z}}$ (omitting the z-dependence) with

$$B_I(x, y) = \sum_{n=-\infty}^{\infty} B_n e^{i\beta_n x} e^{i\kappa_n^{(1)} y} \tag{2.7}$$

$$B_{II}(x, y) = \sum_{n=-\infty}^{\infty} e^{i\beta_n x} \left[C_n^+ e^{i\kappa_n^{(2)} y} + C_n^- e^{i\kappa_n^{(2)} y} \right] \tag{2.8}$$

$$B_{III}(x, y) = \sum_{n=0}^{\infty} A_n \cos\left[\frac{n\pi}{a} \left(x - \frac{a}{2} \right) \right] \cos\left[\kappa_n^{(3)}(y + h) \right], \tag{2.9}$$

and the respective electric fields are expressed as follows:

$$\mathbf{E}_\lambda(\mathbf{r}, t) = \left(\frac{ic^2}{\omega\varepsilon_\lambda} \right) \nabla \times \mathbf{B}_\lambda(\mathbf{r}, t). \tag{2.10}$$

It is worthwhile noting that Eqs.(2.7)–(2.9) were presented explicitly to ensure the following:

- the fields in region I do not diverge as $y \to \infty$; and

- the tangential component of the electric field in region III always vanishes in the surface of the metal.

The mentioned above takes place due to the facts that a perfect metal does not admit non-null electric fields on its inside, and the tangential electric field is continuous across any interface. The coefficients A_n, B_n, and $C_n^\pm$ remain unknown. On the contrary, the electromagnetic wave equation imposes that $\kappa_n^{(1)} = \kappa\,(\beta_n, \varepsilon_1)$, $\kappa_n^{(2)} = \kappa\,(\beta_n, \varepsilon_2)$, and $\kappa_n^{(3)} = \kappa\,(n\pi/a, \varepsilon_3)$, with $\kappa(q, \varepsilon) = \sqrt{\varepsilon\omega^2/c^2 - q^2}$. Moreover, the corresponding fields must obey the Bloch's theorem, $\mathbf{B}(\mathbf{r} + d\hat{\mathbf{x}}) = e^{iqd}\mathbf{B}(\mathbf{r})$ since the grooved system is periodic. The former means that $\beta_n = q + 2n\pi/d$ (q being the momentum of the plasmons in the x-direction). Another consequence of Bloch's theorem lies in the fact that only determination of the fields in an unit cell $|x| < d/2$ of the system is needed. The fields elsewhere are then totally determined.

The evaluation of the boundary conditions at the interfaces I/II and II/III is needed aiming to find the coefficients A_n, B_n, and $C_n^\pm$. It has been shown that the dispersion relation of the spoof plasmons in this system is given by the matrix equation $\det(\mathbf{M} - \mathbf{I}) = 0$, where $\mathbf{I}$ is the unit matrix with elements $|\mathbf{I}|_{lm} = \delta_{lm}$ and $\mathbf{M}$ is a matrix with elements $|\mathbf{M}|_{lm} = M_{lm}$ given by

$$M_{lm} = i\frac{\varepsilon_2 a}{\varepsilon_3 d} \left(\frac{2}{1 + \delta_{l0}} \right) \sum_{n=-\infty}^{\infty} \left(\frac{\chi_n^+ + \chi_n^-}{\chi_n^+ - \chi_n^-} \right) \frac{\kappa_m^{(3)} \sin\left(\kappa_m^{(3)} h \right)}{\kappa_n^{(2)} \cos\left(\kappa_l^{(3)} h \right)} S_{nl}^* S_{nm}, \tag{2.11}$$

with

$$\chi_n^{\pm} = \frac{1}{2}\left[1 + \frac{\sigma\left(\beta_n, \omega\right)\kappa_n^{(1)}}{\omega\varepsilon_0\varepsilon_1} \pm \frac{\varepsilon_2\kappa_n^{(1)}}{\varepsilon_1\kappa_n^{(2)}}\right]e^{i\kappa_n^{(1)}s}e^{\mp i\kappa_n^{(2)}s},$$

and $S_{ln} = \frac{1}{a}\int_{-a/2}^{a/2}dx e^{-i\beta_l x}\cos\left[\frac{n\pi}{a}\left(x - \frac{a}{2}\right)\right]$ in an integral with an analytical solution. In the previous expression, $\sigma\left(q, \omega\right)$ is the non-local conductivity of the graphene sheet (consult the SI for further details). It is worthwhile mentioning that the impact of the graphene to the dispersion is characterized by the factors $\left(\chi_n^+ + \chi_n^-\right)/\left(\chi_n^+ - \chi_n^-\right)$ in each term of the sum, which are equal to 1 in its absence.

At this point, it is interesting to observe that a simple approximation aiming to simplify our calculations includes consideration of only the lowest mode $n = 0$ being non-zero (that is, $A_n = A_0\delta_{n0}$) in region III (the grooves). Thus, the corresponding equation for the dispersion relation of the spoof plasmons is as follows:

$$1 = i\frac{\varepsilon_2}{\varepsilon_3}\frac{a}{d}\frac{\sqrt{\varepsilon_3}\omega}{c}\tan\left(\frac{\sqrt{\varepsilon_3}\omega}{c}h\right)\sum_{n=-\infty}^{\infty}\left(\frac{\chi_n^+ + \chi_n^-}{\chi_n^+ - \chi_n^-}\right)\frac{|S_{n0}|^2}{\kappa_n^{(2)}}, \tag{2.12}$$

where we have used the explicit expression for $\kappa_0^{(3)}$. It can be concluded from Eq. (2.12) that, in the absence of graphene and dispersive dielectric media, it can only have solutions when the momentum of the leading mode in region II, $\kappa_0^{(2)}$, is imaginary, because otherwise the RHS would be imaginary while the LHS is real. The former considerations lead to the fact that the plasmonic solutions are only allowed in the region $q > \sqrt{\varepsilon_2}\omega/c$. This is not surprising and is caused by the bound nature of surface modes. Moreover, the above equation can only have a solution when the tangent function present on its RHS is positive. This leads to the solutions appearing only below a certain maximum frequency $\omega_{\max}$ given by

$$\omega_{\max} = \frac{\pi c}{2h\sqrt{\varepsilon_3}}. \tag{2.13}$$

Comparing this frequency with the effective plasma frequency, it might be concluded that these are exactly the same. It means that the solution is recovered from the effective permittivity model with a completely different model, when the same approximation is applied. It is worthwhile noting that this expression has some limitations. It is only valid for frequency-independent permittivities, but nonetheless it is very useful to make a rough estimation of the order of magnitude of the plasmons frequency, although it tends to overestimate it; following the analogy to the perfect metal/dielectric interface, a better definition of a reference value for the spoof plasmons' fundamental mode frequency is $\omega_{ref} = \omega_{\max}/\sqrt{1 + \varepsilon_1}$.

Apart from allowing the determination of $\omega_{\max}$, Eq. (2.3) has the additional advantage of being much easier to solve than the exact equation $\det(M - I) = 0$, which gets increasingly demanding with the number of modes we introduce in the fields description in region III. Although this approximation looks somewhat I, it will be demonstrated later on that it produces really accurate results.

2.3.3   OPTICAL PROPERTIES

The formalism described in the previous section allows for the calculation of the optical properties of the system (namely its reflectance) in the case of addition of an impinging field in the description of the fields in region I [Eq. (2.8)]. From a practical perspective, this result is particularly useful due to the fact that the reflectance of the system is easily measurable. Moreover, the indirect measurement of other quantities like the absorbance spectrum and the identification of plasmonic resonances (note that there is no transmittance in this system, since a perfect metal is a perfect reflector) is also allowed.

Making an assumption that the impinging electromagnetic wave has a frequency ω and makes some angle θ with the y-axis, the field in region I must be written as follows:

$$B_I(x, y) = B_0 \sum_{n=-\infty}^{\infty} e^{i\beta_n x} \left[e^{-ik_y y} \delta_{n0} + r_n e^{i\kappa_n^{(1)} y} \right], \tag{2.14}$$

where B_0 being the intensity of the impinging magnetic field, $B_n \equiv B_0 r_n$, and r_n being the reflectance amplitudes. In the previous expression, k_y is the momentum of the impinging wave in the y-direction, given $k_y = k \cos(\theta)$, $k = \sqrt{\varepsilon_1} \omega / c$. On the other hand, Bloch's theorem now imposes that $\beta_n = k_x + 2n\pi/d$, $k_x = k \sin(\theta)$. From this point onward, the procedure is completely analogous to the previous one. Following the already described steps (with the differences noted in the SI), we arrive at another matrix equation, this time with the form $(M - I) \cdot A = F$, where the matrices M and I are the same as before, A is a column with elements $|A|_l = A_l$ and F is a column with elements $|F|_l = \phi_l$.

Unlike the previous case (where we found the solution $(M - I) \cdot A = 0$, with no source term), the source introduced by the impinging wave allows the immediate determination of the coefficients r_n, upon the resolution of the previous equation. From these coefficients, the reflectance of the system is simply defined as

$$\Re(\omega) = \sum_{n \in PM} \mathrm{Re} \left\{ \frac{\kappa_n^{(1)}}{\varepsilon_1} \right\} \mathrm{Re} \left\{ \frac{\varepsilon_1}{\kappa_0^{(1)}} \right\} |r_n|^2, \tag{2.15}$$

where $\mathrm{Re}\{x\}$ stands for the real part of x, and both summations are performed strictly over the propagating modes (PM). For the energy scale of interest in this problem (up to a few THz), typically only the fundamental $n = 0$ mode is propagating, and hence the reactance takes the simpler form $\Re(\omega) = |r_0|^2$. The absorbance, on the other hand, corresponds to the fraction of energy which is not reflected, being thus given by $A = 1 - \Re$.

The determination of the r_n coefficients has the additional advantage of allowing the representation of the loss function of this problem, defined as $L(\omega, q) \equiv -\sum_n \mathrm{Im}\{r_n\}$, besides enabling the calculation of the reflectance and absorbance spectra. It is worthwhile noting, that $\mathrm{Im}\{x\}$ stands for the imaginary part of x. All the mentioned above allows for the indirect determination of the dispersion relation of the spoof plasmons even without the approximation

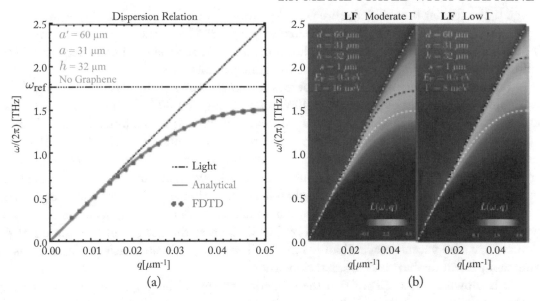

(a)　　　　　　　　　　　　　(b)

Figure 2.17: (a) Comparison between the dispersion relation of the spoof plasmons calculated through Eq. (2.12) (blue) and through an FDTD method (dotted red, retrieved from [100]), in the absence of graphene. (b) Loss function (LF) of the system in the presence of graphene, overlaid by the analytical dispersion relation of the same system with (dashed black) and without (dashed white) graphene, for two different values of the graphene damping energy Γ. All the parameters are specified in each plot. The dot-dashed line in each plot is the dispersion of the light [132].

considered in the previous section. A slightly different definition for the loss function will be used, i.e., $\tilde{L} = \mathrm{sgn}(L)\log(1 + |L|)$. The former is aimed at highlighting the dimmer dispersion curves in the presence of brighter ones. In the previous definition, $\mathrm{sgn}(x)$ stands for the sign of x, and $|x|$ stands for its absolute value.

Graphene's conductivity has been calculated through Mermin's non-local formula [96, 97] synthetically presented in the SI aiming to generate the results presented herein. It is worthwhile mentioning that the main outcomes outlined in the previous section are for isotropic media only. Herein, hBN being anisotropic is considered. The generalization of the previous results to this case is discussed in the SI as well. To conclude, hBN's dielectric function was retrieved from [23].

It is demonstrated in Fig. 2.17b that the dispersion relation of the spoof plasmons can be tuned by the inclusion of a graphene sheet in the system. Doing so, two additional parameters are added for the consideration, i.e., the graphene's Fermi energy E_F and the spacer width s. The former can be easily engineered aiming to control the energy of the spoof plasmons. It is worthwhile mentioning that the qualitative characteristics of their dispersion are kept unchanged. The former tunability feature is demonstrated in Fig. 2.18a–f. The dispersion relation of the spoof plasmons is dramatically affected varying individually the graphene's Fermi energy E_F, the spacer width s, the graphene's relaxation energy Γ, and the maximum frequency in the Brillouin zone (at $q = \pi/d$) in function of the same parameters.

It is confirmed by Fig. 2.18a that there is an actual scaling of the plasmons' energy due to the graphene, which can be higher than 40%, for a doping up to 0.5 eV. Increase of E_F causes the dramatic increase of this enhancement. Moreover, its limit is settled by experimental limitations. Overall, it is difficult to achieve Fermi energies much greater than 0.5 eV [101, 102]. It is worthwhile mentioning, that this enhancement is strongly affected by small spacer widths [see Fig. 2.18b]. The former results from a stronger coupling between the metal and the graphene under these conditions. On the other hand, when the distance between the metal and the graphene increases considerably, the decouple phenomenon takes place and "no graphene" behavior is recovered. Nonetheless, it is demonstrated by Fig. 2.18e that there is a slight saturation of the enhancement for spacers smaller than $\sim 0.5\ \mu m$. Therefore, one can conclude that there is no significant gain in further reducing that dimension.

It is shown in Figs. 2.18c,f that the relaxation energy of the graphene sheet plays a very important role in this analysis. Although this is not an actively changeable parameter, it should be relatively small aiming to guarantee greater energy enhancements. It is worthwhile noting that for Γ values larger than ~ 15 meV, the enhancement is very small even for high graphene doping and small spacer widths. The value of Γ can be significantly reduced employing hexagonal Boron Nitride as a spacer between graphene and the grating [103].

The described feature provides a fertile ground for actively controllable plasmonic waveguides in the THz spectral range. It is possible to excite the spoof plasmons in a grooved surface by means of a system depicted in Fig. 2.19. The former takes advantage of attenuated total reflection (ATR) method [104] to overcome the momentum mismatch between the impinging light and the bound surface modes (similarly to the well-known Kretschmann-Raether [2] or Otto [105] configurations).

Afterward, the energy of these plasmons can be tuned within a reasonable range by applying a gate voltage between the graphene sheet and the metal. Another advantage of using graphene is that its losses are very small. Estimating the propagation length as $\varsigma = 2\pi \upsilon/\omega''$ ($\upsilon = \vartheta\omega'/\vartheta q$ is the group velocity of the spoof plasmons, with $\omega = \omega' - i\omega''$), it is predicted by the calculations that ζ is of the order of millimetres even when the graphene is highly doped ($E_F \geq 0.5$ eV). However, it should be noted that model does not account for the damping in the metal itself. The former does not appear to be a strong limitation, as in the THz the skin depth in the metal is very small, and hence losses in the metal will be small.

In conclusion, it should be noted that the energy-enhanced spoof plasmons studied thus far are effectively hybrid modes that result from the coupling between the plasmons in the metal and in the graphene. Consequently, the strong enhancements observed in Fig. 2.18 are observable in the THz spectral range. Also, they cannot be reproduced for much higher (or lower)

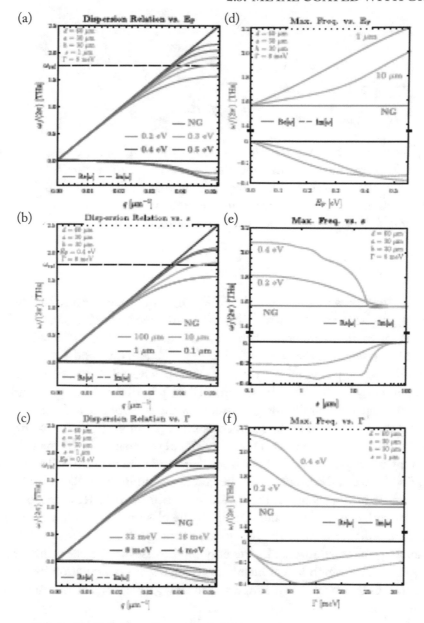

Figure 2.18: Top: dispersion relation of the spoof plasmons for several values of (a) the graphene's Fermi energy E_F, (b) the spacer width s, and (c) the graphene's relaxation energy Γ. Bottom: maximum frequency of the plasmons (for $q = \pi/d$) in function of (a) E_F, (b) s, and (c) Γ. All the parameters are specified in each plot. The dot-dashed line in the left-side plots is the dispersion of the light. "NG" stands for "no graphene" [132].

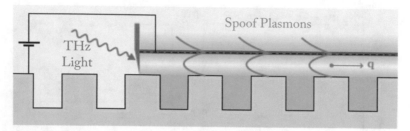

Figure 2.19: Example of a configuration that allows the excitation of spoof plasmons in the system. THz light is impinged in a thin metallic plate which only transmits evanescent modes. The momenta of the evanescent waves is higher than that of the incident light, being able to match the spoof plasmons momenta and thus excite them. The Fermi energy of the graphene sheet can be regulated by applying a variable gate potential [132].

energies. The fact that outside this spectral (THz) range, the characteristic energy scales of the plasmons on the graphene and the metal become very different, and they cannot efficiently couple stands for as the main reason of this behavior. This feature is clearly visible in Fig. 2.20. The loss functions of a system like the one studied in Fig. 2.18 and one whose dimensions were reduced around 30-fold are plotted side-by-side. Doing so, a 30-fold increase in the frequency of the SSPs to around $\omega_{ref} \sim 50$ THz is predicted, well deep in the mid-infrared (mid-IR) region. In both plots, the dispersion of the spoof plasmons without graphene is depicted by the yellow dashed line, and the dispersions of the graphene surface plasmons (GSPs) in a air/graphene/air configuration (this configuration is used for simplicity of the analysis)—by the red dashed lines.

One might conclude, in the THz range, these curves are very close. The former leads to a strong coupling between the plasmons in the graphene and in the metal, and thus provokes a strong enhancement of the energy of the hybrid mode. In the mid-IR range, on the contrary, these curves have different energy scales. Doing so, the plasmons do not couple efficiently, thus provoking a small enhancement of the hybrid mode energy. It is also worthwhile mentioning that, in the latter case, an additional low-energy mode arises in the spectrum, corresponding to a acoustic graphene plasmon in a at metal/air/graphene/air configuration [96] (black dashed line in the figure). One may clearly distinguish the Bragg reflections of this mode at the edges of the Brillouin zone.

Figures 2.20c,d stand for as the another evidence of this feature, with the electric field intensity plotted inside and in a vicinity of a groove for both systems above studied, and both in the presence and absence of graphene. The field distribution in the overall system (including inside the grooves) is dramatically influenced by the inclusion of the graphene into the system (Fig. 2.20c). Thus, the strong enhancement of the field in the groove wedges is observed. Some field oscillations arisen due to acoustic graphene plasmons are observed in Fig. 2.20d.

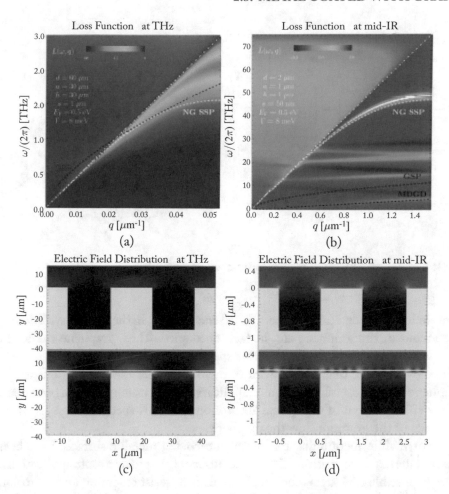

Figure 2.20: Top: Loss function spectrum of two different systems whose spoof plasmons lie on the (a) THz and (b) mid-IR spectral range, in the presence of doped graphene. Overlaid to the loss function are the dispersion of the spoof plasmons in the absence of graphene ("NG SSP," dashed yellow) and the dispersion of the graphene surface plasmons ("GSP," dashed red) in a air/graphene/air configuration. In (b) is also plotted the dispersion of the acoustic plasmons in a at metal/air/graphene/air configuration ("MDGD," dashed black). The white dot-dashed curve is the dispersion of the light in the air. Bottom: electric field intensity in the vicinity of a groove for (c) the THz and (d) the mid-IR corresponding systems, with and without graphene. Both distributions were calculated for $q = 0.9\pi/d$ (for the corresponding d in each system), what translates into the frequencies (c) 1:94 THz and 2:86 THz and (d) 46:3 THz and 47:3 THz, respectively, without and with graphene. All remaining parameters are the same as disclosed in (a) and (b). The color scale is the same in both panels of each figure (c) or (d) [132].

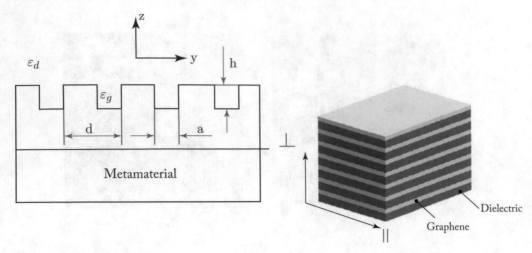

Figure 2.21: Geometry of structured metamaterial surface.

It might be concluded from the performed analysis that in the smaller system the plasmons in the metal and in the graphene are, in fact, decoupled. Thus, graphene cannot be used to efficiently tune the energy of the SSPs in this spectral range.

Although the issue that the behavior of the SSPs under these conditions is not changed by the graphene may be useful for different applications. For example, it has been demonstrated that a strong enhancement of the optical absorption in the IR spectral range [106] is provoked by the presence of the grating below the graphene.

It is worthwhile mentioning that graphene-coated metallic surfaces have been studied recently as an alternative to traditional metallic surfaces for optoelectronic applications. The former feature is possible due to its much higher ability to resist oxidation and corrosion (what is a recurring problem in plasmonics), while keeping (or even improving) the surface's characteristics.

2.4   BOUNDARY OF GRAPHENE-BASED METAMATERIAL AND CORRUGATED METAL

The structure under consideration (Fig. 2.21) consists of two building blocks: grating made of Si ($\varepsilon = \varepsilon_g = 12.25$) and a slab of graphene-dielectric metamaterial. The function of the first building block is creation of higher-order transmission channel(s). In such a case, the coupling of the incident waves differs for the two opposite interfaces. The main its function is to attain tunability by changing the state from dielectric to ENZ, and then to plasmonic one, by engineering μ [107].

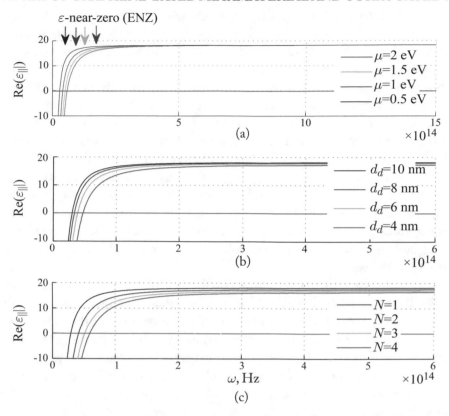

Figure 2.22: The influence of (a), Fermi energy μ, (b) thickness of dielectric d_{md}, and (c) number of graphene sheets N on the real part of $\varepsilon_{\parallel}$. $N = 1, d_{md} = 10$ nm in (a), $N = 1, \mu = 0.5$ eV in (b), and $d_{md} = 10, \mu = 0.5$ eV in (c).

The effective-medium approach is applied aiming to describe the optical response of such a system. The former is justified if the wavelength of the radiation considered is much larger than the thickness of any layer. It is based on averaging the structure parameters. Hence, further in this paper we consider the effective homogeneous media for the semi-infinite periodic structures. The effective permittivities are as follows [108]:

$$\varepsilon_{\parallel}^{m} = \frac{\varepsilon_{mg} d_{mg} + \varepsilon_{md} d_{md}}{d_{mg} + d_{md}} \tag{2.16}$$

$$\varepsilon_{\perp}^{m} = \frac{\varepsilon_{mg} \varepsilon_{md} \left(d_{mg} + d_{md}\right)}{\varepsilon_{mg} d_{md} + \varepsilon_{md} d_{mg}}, \tag{2.17}$$

where ε_{mg}, ε_{md} − are the permittivities of the graphene and dielectric layers correspondingly; d_{mg}, d_{md} − are the thicknesses of the graphene and dielectric layers correspondingly.

Matching the tangential components of the electrical and magnetic fields at the interface implies the dispersion relation for the surface modes localized at the boundary separating two anisotropic media [109]. We assume the permittivity $\varepsilon_{mg}(\omega)$ to be frequency dependent as the corresponding layer is represented by graphene.

Within the random-phase approximation and without an external magnetic field, graphene may be regarded as isotropic and the surface conductivity can be written as follows [110, 111], $\sigma = ie^2\mu/\pi h^2(\omega + i/\tau)$, where ω, h, e, μ, τ is the frequency, Planck constant, charge of an electron, chemical potential (Fermi energy), and phenomenological scattering rate, respectively. The Fermi energy μ can be straightforwardly obtained from the carrier density n_{2D} in a graphene sheet, $\mu = hv_F\sqrt{\pi n_{2D}}$, v_F is the Fermi velocity of electrons. It should be mentioned that one can electrically control the carrier density n_{2D} by an applied gate voltage, thus leading to a voltage-controlled Fermi energy μ. Here we assumed that the electronic band structure of a graphene sheet is unaffected by the neighboring layers. Thus, the effective permittivity ε_{mg} of graphene can be calculated as follows [112]: $\varepsilon_{mg} = 1 + i\sigma/\varepsilon_0\omega d_{mg}$, where ε_0 is the permittivity in the vacuum.

One can see that $\mathrm{Re}\left(\varepsilon_{\parallel}\right)$ crosses zero at a frequency which depends on μ. For instance, this happens at 30 THz when $\mu = 0.5$ eV, and at 45 THz when $\mu = 1$ eV. Consequently, spectral location of the region of transition from the effectively plasmonic to the effectively dielectric state can be significantly shifted by means of variations in μ. Note that the gate positioning will be considered at the next steps. Generally, electrical gating of a multilayer graphene-dielectric metamaterial is a challenging task. One of possible gating schemes is presented in Ref. [109]. In Ref. [113], chemical doping has been used while preparing each graphene layer using CVD, instead of electrical gating. In Ref. [113], the double-layer graphene has been experimentally gated, and the method was presented by the authors as the one being usable for the structures composed of a larger number of graphene monolayers. So, from a practical point of view, this approach can be utilized also in the case of graphene-dielectric metamaterial containing finite number of graphene layers.

In order to find the dielectric parameters of the effective medium describing gratings, consider a periodic assembly of parallel plates. The effective dielectric constants of such an assembly are as follows [68]:

$$\varepsilon_x = \varepsilon_z = \frac{(d-a)\varepsilon_{Ag} + a\varepsilon_g}{d} \tag{2.18}$$

$$\varepsilon_y = \frac{d}{(d-a)/\varepsilon_{Ag} + a/\varepsilon_g}; \tag{2.19}$$

where ε_d is the permittivity of the surrounding media and ε_{Ag} – is the permittivity of silver.

Herein, we present a simple example of polaritons in corrugated metal at THz frequencies. The wave vector k [115] is plotted as a function of the frequency aiming to illustrate the properties of SPPs. We deal with Ag case [116]. It is assumed that the structure is surrounded by silicon, i.e., Si ($\varepsilon = \varepsilon_g = 12.25$).

Figure 2.23a depicts the dispersion curves of SPPs at the boundary metamaterial/structured surface with $d = 10$ nm. It is of particular importance to investigate the impact of the period on the dispersion curves of SPPs. In doing so, three different groove widths a are considered. One may conclude from Fig. 2.23 that the asymptotic frequency decreases with an increase in groove width. Figure 2.23b presents the losses of these SPPs as a function of frequency. It is worthwhile noting that the loss of SPPs is dramatically influenced by the increase in frequency. Besides that, it is of particular importance to note, that the forbidden region between the modes starts squeezing as presented in the Fig. 2.23a and the approach each other with the decrease of the groove width.

It is of particular interest to analyze the effect of the lattice constant (d) on the dispersion of SPPs. The dispersion curves for SPPs at the boundary metamaterial/corrugated surface with different lattice constants $d = 5, 7, 10$ nm, respectively, are displayed in Fig. 2.24a. The groove parameter is $a = 2$ nm for all cases. Figure 2.24b shows the losses of SPPs for three cases. It can be concluded from Fig. 2.24b that a larger loss of SPPs for a given frequency takes place in case of a smaller lattice constant. One might conclude from Fig. 2.24b that losses of the SPPs are drastically influenced by the lattice constant. For a given frequency, an increase of the lattice constant may cause a significant reduction of the loss of spoof SPPs. A low-loss THz waveguiding system is essential because of the need for a compact, reliable, and flexible THz system for various applications.

The impact of the chemical potential on the propagating modes along metamaterial/corrugated metal interface is depicted in Fig. 2.25. One may engineer the chemical potential μ of graphene by tuning the gate voltage or doping [112]. It is worthwhile noting that the chemical potential has different values in these dispersion curves, i.e., $\mu = 0.1$ eV, $\mu = 0.5$ eV, $\mu = 1$ eV, and $\mu = 1.5$ eV. It has been confirmed that the chemical potential has a dramatic impact modulating the modes. It should be noted that the resonance frequency for the modes increases with the increase of chemical potential.

Figure 2.26 displays the propagation length $L_p = \frac{1}{2\mathrm{Im}(\beta)}$ as a function of the incidence terahertz frequency ($\omega = 0.1 - 500$ THz) for different values of groove width ($a = 0.4d$; $a = 0.6d$) and chemical potential ($\mu = 0.1$ eV, 0.5 eV, 1 eV, 1.5 eV). It is obvious that groove width (a) and the chemical potential (μ) play a significant role in modulation of the propagation length of the modes. The propagation length of the modes increases increasing groove width and chemical potential. It is of particular interest to mention that the propagation length profiles of the modes follow the exponential decay. The usage of graphene allows for the increase the propagation length of the modes compared to ordinary interface modes.

2.5 SPOOF SURFACE PLASMONS IN SPOOF–INSULATOR–SPOOF WAVEGUIDES

The problem under analysis is sketched in Fig. 2.27, defining all the geometrical elements. Thus, the specific system, for which all the effects will be illustrated, consists of two structures with

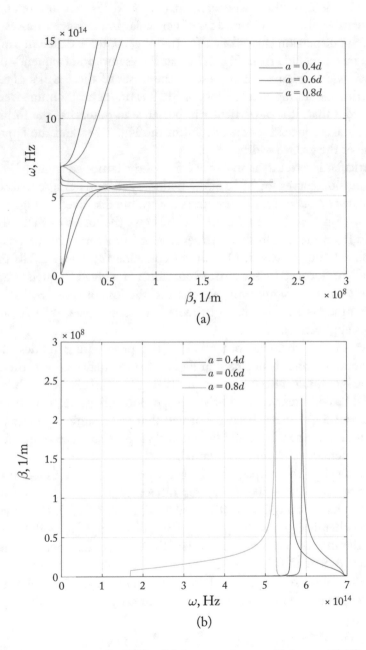

Figure 2.23: (a) Dispersion curves for SPPs. (b) Attenuation coefficients of SPPs, lattice constant $d = 10$ nm.

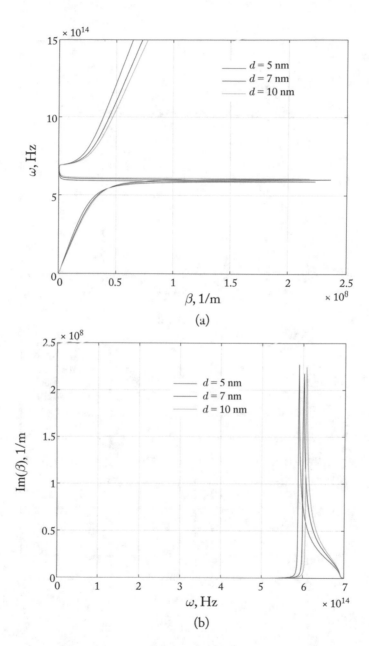

Figure 2.24: Dispersion curves (a) and attenuation coefficients (b) of SPPs for different lattice constants $d = 5, 7, 10$ nm, respectively. Parameters of grooves: $a = 2$ nm.

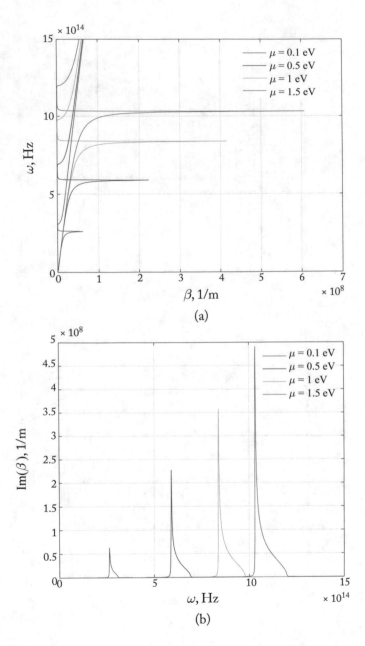

Figure 2.25: Influence of chemical potential on the dispersion curves (a) and attenuation coefficients (b) of SPPs at nanostructure metamaterial interface with $d = 10$ nm, $a = 0.4d$.

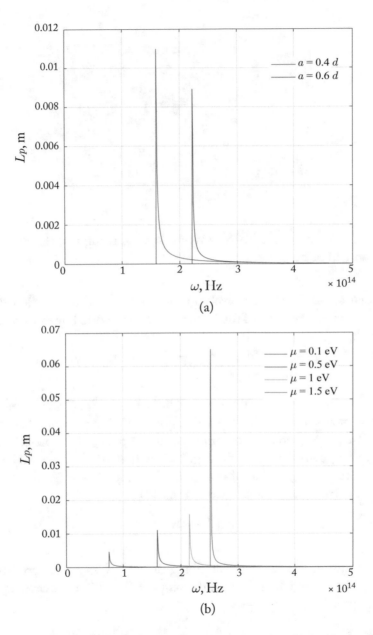

Figure 2.26: Influence of groove width (a) and chemical potential (b) on the propagation length of surface wave modes as function of incidence frequency.

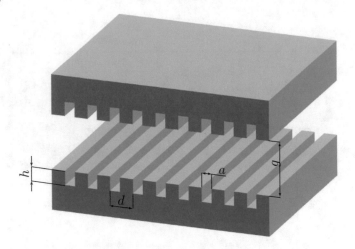

Figure 2.27: Spoof-insulator-spoof (SIS) waveguide comprising two counter-facing structures separated by an air gap of height g.

a periodic corrugation of deeply-subwavelength grooves of width a, period d, and height h. Specifically, the dispersion relation of the surface modes supported by the structure of Fig. 2.27 is as follows [117]:

$$\frac{k_z^g}{k_0} = \frac{\varepsilon_d}{\varepsilon_s} \tan\left(k_0 \sqrt{\varepsilon_d} h\right) \tanh^{\pm 1}\left(k_z g\right), \tag{2.20}$$

where $k_0 = 2\pi/\lambda_0$, λ_0 is the free space wavelength, $k_z^g = i k_z^g = \sqrt{\beta^2 - k_0^2 \varepsilon_d}$, k_z is the z-component of the wavevector in the gap and k_z^g is used to represent the evanescent quality of the wave, β is the propagation constant in the x-direction, ε_d is the dielectric function of the bounding dielectric, h, g are defined in the Fig. 2.27, with $d \ll \lambda_0$. It should be mentioned that $\pm$ corresponds to the bonding and antibonding modes, respectively.

An effective permittivity ε_s is defined as follows [77]:

$$\varepsilon_s = \frac{(d-a)\varepsilon_c + a\varepsilon_g}{d}. \tag{2.21}$$

Here, ε_g is the permittivity of the material, filling the groves, ε_c is the permittivity of the corrugated material, which strongly influences the outputs. Two different cases will be considered throughout the paper.

Spoof Surface Plasmon Modes
In this section we will focus on the examination of the designer spoof surface plasmon modes in two different structures, i.e., semiconductor SIS, SIS made of TCO.

Semiconductor Case

A case in point is the SIS structure made up of semiconductor. The example of a heavy-doped Si is considered. It is assumed, that the permittivity of the semiconductor is defined by the Drude model, i.e., $\varepsilon(\omega) = \varepsilon_\infty \left(1 - \frac{\omega_p^2}{\omega^2}\right)$ with ε_∞ being the background permittivity and ω_p the plasma frequency.

In Fig. 2.28, we sketch the dispersion relation for a representative SIS structure employed to operate in the THz regime, where spoof plasmonic structures have been proposed and experimentally demonstrated [118, 119]. The geometric parameters are as follows: $g = 0.4\ \mu$m, $d = 0.5\ \mu$m. The range of plotted propagation constants β is significantly below the Brillouin zone boundary at $\beta_{BZ} = \pi/d$, confirming the validity of the metamaterial limit and Eq. 2.20. It is at the center of an unprecedented interest to analyze the effect of the groove depth on the dispersion curves of spoof SPPs. For this reason, three different groove depths, i.e., $h = 3d, h = 2.5d$, and $h = 1.5d$ are suggested for the study. It should be pointed out that several values of the groove width are considered for each groove depth. As seen from Fig. 2.28, the larger groove depth corresponds to a lower asymptotic frequency. This is also the case for the groove width. However, the groove depth drastically affects asymptotic frequency compared to the groove width. This situation is likely to be realized dealing with the periodically corrugated semiconductor structure [77]. For the sake of comparison, we also plot the modes for the metal case in Fig. 2.28a, assuming that $\varepsilon_s = d/a$ [117]. The doubly corrugated metal surfaces have already been studied in [120]. The plotted dispersion modes [120] differ from the results obtained in Fig. 2.28a due to the difference in the geometrical dimensions. It should be mentioned that modes in the SIS waveguide geometry reveal strong dispersion and approach the asymptote as $\beta \to \infty$. Since we can no longer treat the corrugated layer in the effective medium limit when the wavelength in the structure approaches the periodicity of the corrugations, we restrict ourselves and consider that Eq. 2.20 is valid only for the modes when β is less than π/d by approximately a factor of two. Additionally, in the current analysis we ignored the losses that are present in systems involving real metals [117]. In other words, the periodicity of the structure presents a band edge at $\beta_{edge} = \pi/d$ initiating the effective material assumption to break down [24]. Moreover, we note that our calculations also show a negligible dependence on the geometrical parameter a. The former issue in turn confirms that the structures are in the effective medium limit.

Transparent Conducting Oxides Case

The case of ZnO (AZO) is considered [123]. The permittivity of the transparent conducting oxide, that is an essential property of any plasmonic material, is defined by the Drude–Lorentz model, i.e.,

$$\varepsilon_{tco}(\omega) = \varepsilon_b - \frac{\omega_p^2}{\omega\left(\omega + i\gamma_p\right)} + \frac{f_1\omega_1^2}{\left(\omega_1^2 - \omega^2 - i\omega\gamma_1\right)},$$

where the values of the parameters are listed in Table 2.1 [122].

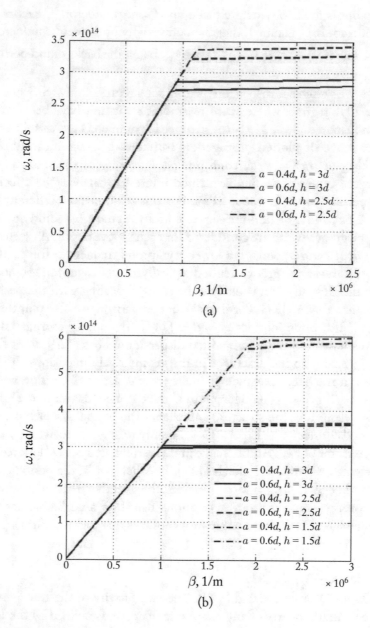

Figure 2.28: Bonding modes in a SIS metal (a) and Si (b) waveguides for $g = 0.4$ μm, $d = 0.5$ μm.

Table 2.1: Drude–Lorentz parameters of the alternative plasmonic material retricved from ellipsometry measurements [122]

	AZO
εb	3.5402
$\omega_p[eV]$	1.7473
$y_p[eV]$	0.04486
f_1	0.5095
$\omega_1[eV]$	4.2942
$y_1[eV]$	0.1017

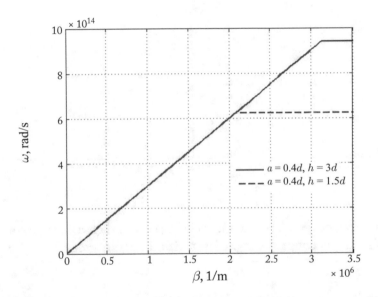

Figure 2.29: Bonding modes in a SIS AZO waveguide for $g = 0.4\ \mu$m, $d = 0.5\ \mu$m.

The effect of the groove depth on the dispersion curves of spoof SPPs is illustrated in Fig. 2.29. For this reason two different groove depths, i.e., $h = 3d$ and $h = 1.5d$ are suggested to evaluate the influence of the geometry. In other words, we show the outputs for various groove depths. As seen from Fig. 2.29, the asymptotic frequency can be effectively decreased by decreasing the groove depth. In other words, we have obtained the inverse phenomenon in comparison with the periodically corrugated transparent conducting oxide structure [121]. It should be mentioned that a real part of the propagation constant depicted in Fig. 2.29 is not influenced by the presence of the imaginary part in the complex permittivity defined by the Drude–Lorentz model. Moreover, to the best of our knowledge, transparent conducting oxides

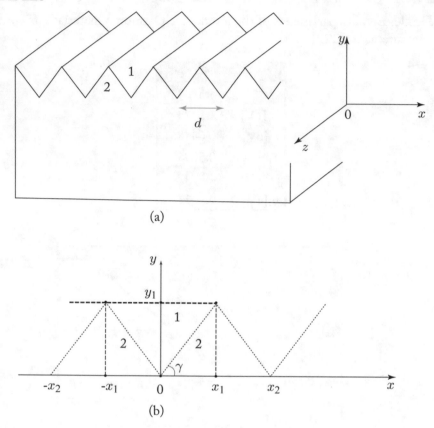

Figure 2.30: (a) Array of grooves drilled in metal placed in dielectric with permittivity ε_1 and (b) geometry of homogeneous anisotropic layer with triangular grooves.

are known as low-loss plasmonic materials [121, 122]. To sum up, the dispersion curves exhibit SPP-like behavior, presenting an asymptote controlled mainly by h.

2.6 SPOOF PLASMONS IN METALS WITH A TRIANGULAR SHAPE OF PERFORATION

Consider a corrugated metal surface, in which an array of grooves is machined. This structure is shown in Fig. 2.30a. The surface is continuous in the z-direction and there are two different material in the $x-y$ plane. Such an interface can be modeled as a three-layer structure, consisting of a homogeneous anisotropic layer of a thickness h—see Fig. 2.30a—describing corrugations, placed between a metal and a dielectric. The approximation of the central layer as an anisotropic effective medium is possible as long as the period d is much smaller that the wavelength of the electromagnetic field.

We thus investigate a two-dimensional (2D) waveguide with triangular grooves supporting an even TM mode with a propagation constant β that satisfies a dispersion relation. The corrugation at interfaces between regions 1 and 2 is periodic with a period d, i.e., the surface can be described by some function $g(x)$ where $g(x) = g(x + d)$. The height of the corrugation is h, as shown in Fig. 2.30a.

We assume propagation in the x-direction and no variation of fields in the z-direction. This makes it a 2D problem. Due to the continuity in the z-axis Maxwell's equations decouple into TE (H_x, H_y, E_z) and TM (E_x, E_y, H_z) polarizations.

Effective Permittivity for Arbitrary Grooves

The challenge here lies in determination the macroscopic property of interest, e.g., permittivity. As long as the size and spacing of the grooves are much smaller than the wavelength, a perforated perfectly conducting surface behaves as an effective medium. This medium is characterized by the effective dielectric constants. In order to find the dielectric parameters of the effective medium, consider a periodic assembly of metallic tilted plates forming a corner.

Suppose that the electric vector of a plane monochromatic wave is perpendicular to the interface separating two regions. The electric displacement must have the same value in both regions [68]. If E_1 and E_2 are the corresponding electric fields, then

$$E_1 = \frac{D_{ny}}{\varepsilon_1}, \tag{2.22}$$

$$E_2 = \frac{D_{ny}}{\varepsilon_2}, \tag{2.23}$$

where $D_{ny} = D \cos \gamma$ with γ being the angle with respect to the boundary—see Fig. 2.30b. Here, ε_1 and ε_2 are the permittivities of the region 1 and 2, denoted in Fig. 2.30, correspondingly. The field E averaged over the total volume is

$$E = \frac{S_1 \dfrac{D \cos \gamma}{\varepsilon_1} + S_2 \dfrac{D \cos \gamma}{\varepsilon_2}}{S_1 + S_2}, \tag{2.24}$$

where S_1 and S_2 are the corresponding areas (see Fig. 2.30b). The effective permittivity $\varepsilon_\perp$ is thus

$$\varepsilon_x = \varepsilon_\perp = \frac{D}{E} = \frac{(S_1 + S_2)\, \varepsilon_1 \varepsilon_2}{(S_1 \varepsilon_2 + S_2 \varepsilon_1) \cos \gamma}. \tag{2.25}$$

Suppose next that the incident field has its electric vector parallel to the interface. The tangential component of the electric vector is continuous across the interface discontinuity [68].

The electric displacement in the two regions therefore is

$$D_1 = \varepsilon_1 E_{\tau x}, \tag{2.26}$$
$$D_2 = \varepsilon_2 E_{\tau x}, \tag{2.27}$$

where $E_{\tau x} = E \cos \gamma$. The average electric displacement is

$$D = \frac{S_1 \varepsilon_1 E \cos \gamma + S_2 \varepsilon_2 E \cos \gamma}{S_1 + S_2}. \tag{2.28}$$

Hence, the effective permittivity is now given by

$$\varepsilon_z = \varepsilon_y = \varepsilon_{\parallel} = \frac{D}{E} = \frac{\cos \gamma \, (S_1 \varepsilon_1 + S_1 \varepsilon_2)}{S_1 + S_2}. \tag{2.29}$$

In the next section we will present the field distributions having in mind the effective medium approximation approach described above.

Field Distribution for Various Grooves

Using the effective medium approximation derived in the previous section, one can describe the transverse magnetic spoof plasmon mode propagating in the structure. The only non-zero components of the magnetic and electric fields expressed in terms of ε_x, ε_y, ε_z are H_z, E_x, E_y [66]:

$$H_z = A e^{-j\beta x} \begin{cases} e^{-k_y y}, & y > g(x), \quad y < 0 \\ \dfrac{\cos \left(k_0 \left(y - y_1 \right) \right)}{\cos \left(k_0 y_1 \right)}, & 0 < y < g(x) \end{cases} \tag{2.30}$$

$$E_x = -\frac{\partial_y H_z}{j k_0 \varepsilon_x} = -\frac{A j e^{-j\beta x}}{k_0} \begin{cases} \dfrac{k_y}{\varepsilon_1} e^{-k_y y}, & y > g(x), \quad y < 0 \\ \dfrac{k_0}{\varepsilon_x} \cdot \dfrac{\sin \left(k_0 \left(y - y_1 \right) \right)}{\cos \left(k_0 y_1 \right)}, & 0 < y < g(x) \end{cases} \tag{2.31}$$

$$E_y = \frac{\partial_x H_z}{j k_0 \varepsilon_y} = -\frac{A \beta e^{-j\beta x}}{k_0} \begin{cases} \dfrac{1}{\varepsilon_1} e^{-k_y y}, & y > g(x), \quad y < 0 \\ \dfrac{\cos \left(k_0 \left(y - y_1 \right) \right)}{\varepsilon_y \cos \left(k_0 y_1 \right)}, & 0 < y < g(x), \end{cases} \tag{2.32}$$

where k_y – is the transversal propagation constant, defined as $k_y^2 = \omega^2 \mu \varepsilon_2 - \beta^2$, $k_0 = 2\pi f/c$ is the vacuum wave vector, f is the frequency, and β – is the longitudinal propagation constant, expressed as $\beta = k_0 \sqrt{\varepsilon_1 + \frac{\varepsilon_1^2}{\varepsilon_x^2} \tan^2 \left(k_0 y_1 \right)}$, A – is the amplitude.

Numerical results presented in this section have been obtained for a triangular shape of the corrugation. Equations (2.30)–(2.32) have been plotted by means of Matlab. The ranges for x and y have been chosen and the values of the field components have been generated over

Table 2.2: The parameters of the structure

Parameter	Value				
ε_1	1				
ε_2	1				
ε_x	52.6347	for	-1.1154	for
ε_y	0.0190	$x_1 = 0.01\,\mu m$	-0.8966	$x_1 = 17\,\mu m$
ε_z	0.0190		-0.8966	

these ranges, producing the 3D plots. It is interesting to notice, that the increments for x and y have been chosen small enough to yield the smooth field distributions. The triangle angle has been considered as a parameter. All parameters of the investigated structure are summarized in Table 2.2. Here x_1 is the half side of the triangle—see Fig. 2.30b. The values of the parameters including the permittivities ε_x, ε_y, ε_z are different for different values of x_1.

The possibility of focusing using a non-perfect metal with periodically drilled grooves has already been considered in [66]. The authors considered a 1D periodic array of rectangular grooves. It was shown that the focus size was limited to tens of micrometers. Considering a structure with periodic array of triangular grooves we have achieved better nanofocusing. To illustrate the nature of the plasmonic confinement, the fields given by Eqs. (2.30)–(2.32) for different dimensions of the grooves are plotted in Figs. 2.31 and 2.32. The origin of the coordinate system in all graphs is shifted by y_1 value with respect to the origin of the coordinate system presented in Fig. 2.30b. For all graphs $y_1 = 30\ \mu$m. The fields are plotted in the plane xy in Fig. 2.31a–c for $x_1 = 0.01\ \mu$m, and for $x_1 = 17\ \mu$m in Fig. 2.32a–c. All the fields have been calculated at frequency $f = 3$ THz.

One can see a dependence $h(x)$ in Figs. 2.31 and 2.32, i.e., an appropriate grading of the groove depth or the thickness of the effective medium. The method of the adiabatic concentration has been developed in [124]. Figures 2.31 and 2.32 demonstrate the effect of increasing the dimension x_1. Our simulations show that the effect of focusing is strongly influenced by the system dimensions. Thus, one can control the location of hot spots by manipulating system dimensions. It should be mentioned that hot spots are highly confined fields. For applications such as addressing of absorbing particles at different positions on the surface, it is of particular interest to dynamically shift the position of the bright hot spot. Such optical hot spots are useful for scanning the probe microscopy, or a plasmonic sensing technique used for optical or chemical mapping of surfaces with sub-diffraction-limited resolution.

As one can see, when the depth of the grooves h is adiabatically increased (shown by the green line), the field strength and its intensity of the spoof plasmon increase. One can decrease the hot spot size by decreasing the sizes of the investigated system. An important observation

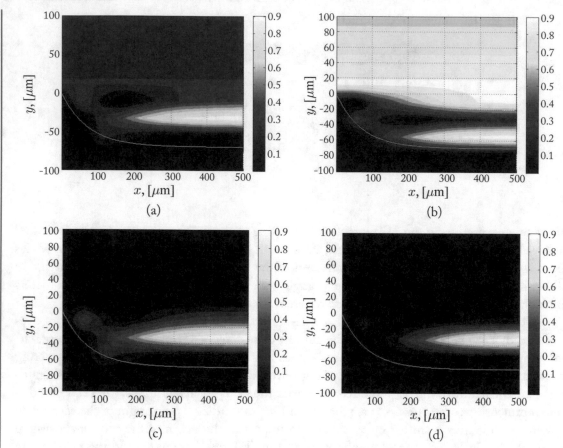

Figure 2.31: Spatial distributions of non-zero components of magnetic and electric fields due to adiabatic increase of thickness of effective medium for $x_1 = 0.01$ μm. The green line represents the shape of the effective medium, i.e., $h(x)$. (a) Transverse magnetic field H_z as function of co-ordinate y; (b) same as (a) but for longitudinal electric field E_x; (c) same as (a) but for transverse electric field E_y; and (d) intensity distribution of spoof plasmons in xy plane as function of x.

is that as x_1 decreases, the intensity of E_y component increases and the intensity of the E_x component decreases.

We have estimated the minimum size of the hot spot that can be achieved by concentration of the spoof plasmons on metals. In the case of a perfect conductor, the size of the hot spot is not limited by the losses and can theoretically be arbitrarily small [66]. This analysis leads to a minimum achievable hot spot size on the order of 10–20 μm. The size of the hot spot was obtained considering its dimension in the x direction. Note that this is for a frequency

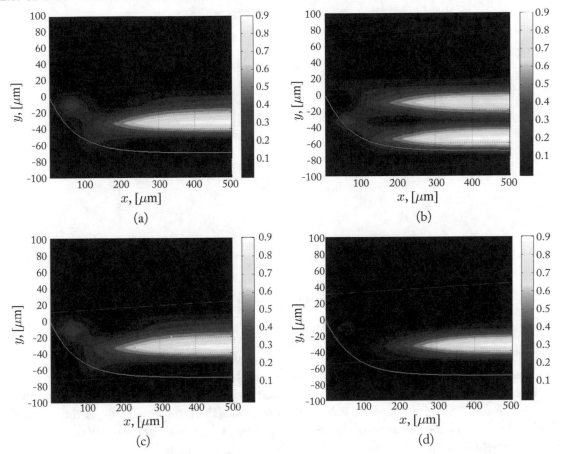

Figure 2.32: Spatial distributions of non-zero components of magnetic and electric fields due to adiabatic increase of thickness of effective medium for $x_1 = 17$ μm. The green line represents the shape of the effective medium, i.e., $h(x)$. (a) Transverse magnetic field H_z as function of co-ordinate y; (b) same as (a) but for longitudinal electric field E_x; (c) same as (a) but for transverse electric field E_y; and (d) intensity distribution of spoof plasmons in xy plane as function of x.

of $f = 3$ THz, i.e., a wavelength of 100 μm. An achievable spot size is thus well beyond the diffraction limit.

The cross-sections of field profiles vs. position have been calculated. The dependencies of the normalized transverse magnetic field for different values of x_1 are depicted in Fig. 2.33.

The extrema points in Fig. 2.33 exist due to the fact that the behavior of the magnetic field is strongly influenced by the function $\cos\left(\beta_g\left(y - y_1\right)\right) / \cos\left(\beta_g y_1\right)$. An enlarged view of the region possessing the extrema points is depicted in the inset of the figure.

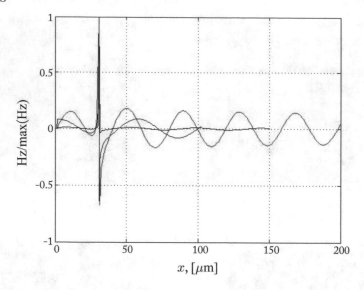

Figure 2.33: Cross-sections of field profiles for different values of x_1: blue $- x_1 = 17\ \mu$m; black $- x_1 = 25\ \mu$m; red $- x_1 = 35\ \mu$m.

One can also observe in Fig. 2.33 that if the value of x_1 increases, the size of the hot spot decreases. This would then suggest that the length of the triangle side is a good degree of freedom for achieving the desirable size of the hot spot. In addition, the increase in the value x_1 leads to a shift of the magnetic field peak amplitude. Thus, the amplitude of the magnetic field shifts due to a modification of the structure geometry. Therefore as shown in the figure, the magnetic field is easily affected by changes of the structure geometry.

CHAPTER 3

Conclusions

A method for analyzing spoof SPPs on a periodically corrugated surface has been presented. Compared to the previous works, our approach enables us to investigate the semiconductor case, since it takes into account the permittivity of the semiconductor expressed by the Drude model. In the THz frequency range, the properties of the dispersion and loss of spoof SPPs on corrugated Si surfaces have been analyzed. The asymptotic frequency of spoof SPPs mainly depends on the groove depth, but the loss of spoof SPPs is sensitive to all parameters of the surface structure, including the lattice constant. However, the performance of low-loss propagation for spoof SPPs can be achieved by an optimum design of the surface structure.

Herein, we have also demonstrated that the widely studied tunability of graphene can be successfully applied to a metallic grating aiming to engineer its dispersion relation. The former feature is especially favorable in the THz spectral range. It has also been demonstrated that this feature can have several applications, ranging from optoelectronic waveguides to filters and to THz sensing, with the additional bonus of providing graphene-protected metallic surfaces.

The model we employed proved to be accurate when benchmarked against FDTD calculations. However, it has some limitations that should be properly emphasized. The most important one is the fact that the metal is considered ideal. The former means that its validity is restricted to situations where the metal's skin depth is negligible, such as the case of the THz and mid-IR radiation in most good plasmonic metals. For the same reason, although this model accounts for non-local effects in the graphene, it cannot account for non-local effects in the metal surface, that should be important when the graphene sheet is very close (a few nanometers) from the metal surface; under those conditions, further corrections need to be employed to ensure experimentally accurate results. However, in our case, the distance between graphene and the metal grating is large enough for non-local effects to negligible compared to that length scale.

The nanostructured metamaterial interface is used to excite the SPP waves. The dispersion relationship is computed by implementing the effective medium theory and transfer matrix approach and the following conclusions can be drawn.

- Surface wave modes propagate along the studied interface.

- One may tune the bandgap corresponding to non-propagation regime by engineering the groove width or chemical potential of graphene.

- The propagation length as a function of the terahertz frequency range is studied. It has been concluded that under appropriate parameters the propagation length can be modulated. The present method of surface wave modulation is quite simple in comparison with the corrugated structures [125].

• The proposed geometry can be used for optical sensing and wave propagation in the terahertz regime.

Herein, we have considered a formalism to analyze SPPs at the metamaterial/corrugated metal interface. In comparison with the previous works, our approach enables us to investigate the metamaterial case since it takes into account the permittivity of the metamaterial expressed by the effective medium theory. We have analyzed the properties of the dispersion and loss of SPPs at the metamaterial/corrugated metal boundary at the THz frequency range. The groove width drastically affects the asymptotic frequency of SPPs. On the other hand, the loss of SPPs is sensitive to all parameters of the surface structure. However, the performance of low-loss propagation for spoof SPPs can be achieved by an optimum design of the surface structure.

We have developed a theory, which describes spoof plasmons propagating on a metal perforated with a 1D periodic array of triangular grooves. We have also considered the possibility of adiabatic concentration of THz energy by spoof plasmons. We have shown that the concentration of spoof plasmons to a spot with the size of 10–20 μm is possible at a wavelength of 100 μm which is well beyond the diffraction limit. The described properties may be useful for plasmonic circuitry applications, surface optical tweezers, optical data storage, and biosensing.

Analysis of the surface plasmon modes dispersion relations for a SIS structure with semiconductor and TCO parameters has been presented in this work. While these spoof surface modes are predicted for perfect conductive gratings, we have shown that this is also the case for the SIS structures made of semiconductor and TCO. As verified by extensive numerical analysis, it is anticipated that these SIS structures will be useful as waveguides due to the extensive range of tenability in their dispersion curves provided by the many geometrical degrees of freedom of the structures, enabling applications such as group-velocity delay lines and photonic-plasmonic mode converters.

Bibliography

[1] R. H. Ritchie, Plasma losses by fast electrons in thin films, *Phys. Rev.*, 106:874, (1957). DOI: 10.1103/physrev.106.874. ix, xi

[2] H. Raether, *Surface Plasmons on Smooth and Rough Surfaces and on Gratings*, Springer-Verlag, Berlin, 1988. DOI: 10.1007/bfb0048317. ix, 30

[3] P. S. Chandrasekhar, N. Chander, O. Anjaneyulu, and V. K. Komaral, Plasmonic effect of Ag@TiO2 core—shell nanocubes on dye-sensitized solar cell performance based on reduced grapheme oxide—TiO2 nanotube composite, *Thin Solid Films*, 594:45–55, (2015). DOI: 10.1016/j.tsf.2015.10.013. ix

[4] E. P. Bellido, Y. Zhang, A. Manjavacas, P. Nordlander, and G. A. Botton, Plasmonic coupling of multipolar edge modes and the formation of gap modes, *ACS Photonics*, 4:1558–1565, (2017). DOI: 10.1021/acsphotonics.7b00348. ix

[5] N. K. Pathak, A. Ji, and R. P. Sharma, Study of efficiency enhancement in layered geometry of excitonic-plasmonic solar cell, *Appl. Phys. A Mater. Sci. Process.*, 115:1445–1450, (2014). DOI: 10.1007/s00339-013-8061-0. ix

[6] J. Butet and O. J. F. Martin, Surface-enhanced hyper-Raman scattering: A new road to the observation of low energy molecular vibrations, *Phys. Chem. C*, 119:15547–15556, (2015). DOI: 10.1021/acs.jpcc.5b04128. ix

[7] N. K. Pathak and R. P. Sharma, Study of broadband tunable properties of surface plasmon resonances of noble metal nanoparticles using mie scattering theory: Plasmonic perovskite interaction, *Plasmonics*, 11:713–719, (2016). DOI: 10.1007/s11468-015-0097-x. ix

[8] F. Rundong, W. Ligang, C. Yihua, Z. Guanhaojie, L. Li, Z. Lia, et al., Tailored Au@TiO2 nanostructures for the plasmonic effect in planar perovskite solar cells, *J. Mater. Chem. A*, 5:12034–12042, (2017). DOI: 10.1039/c7ta02937c. ix

[9] T. Nagatsuma, G. Ducournau, and C. C. Renaud, Advances in terahertz communications accelerated by photonics, *Nat. Photonics*, 10:371–379, (2016). DOI: 10.1038/npho-ton.2016.65. ix

[10] J. F. O'Hara, W. Withayachumnankul, and I. Al-Naib, A review on thin-film sensing with terahertz waves, *J. Infrared Millim. Tera. Waves*, 33:245–291, (2012). DOI: 10.1007/s10762-012-9878-x. ix

[11] X. Yang, X. Zhao, K. Yang, Y. Liu, W. Fu, and Y. Luo, Biomedical applications of terahertz spectroscopy and imaging, *Trends Biotechnol.*, 34:810–824, (2016). DOI: 10.1016/j.tibtech.2016.04.008. ix

[12] H.-B. Liu, H. Zhong, N. Karpowicz, Y. Chen, and X.-C. Zhang, Terahertz spectroscopy and imaging for defense and security applications, *P. IEEE*, 95:1514–1527, (2007). DOI: 10.1109/jproc.2007.898903. ix

[13] T. -I. Jeon and D. Grischkowsky, THz Zenneck surface wave (THz surface plasmon) propagation on a metal sheet, *Appl. Phys. Lett.*, 88:061113, (2006). DOI: 10.1063/1.2171488. ix, xi, xii

[14] D. G. Cooke and P. U. Jepsen, Optical modulation of terahertz pulses in a parallel plate waveguide, *Opt. Express.*, 16:15123–15129, (2008). DOI: 10.1364/oe.16.015123. x

[15] J. Gómez Rivas, M. Kuttge, H. Kurz, P. Haring Bolivar, and J. A. Sánchez-Gil, Low-frequency active surface plasmon optics on semiconductors, *Appl. Phys. Lett.*, 88:082106, (2006). DOI: 10.1063/1.2177348. x

[16] M. Rahm, J.-S. Li, and W. J. Padilla, THz wave modulators: A brief review on different modulation techniques, *J. Infrared Millim. Tera. Waves*, 34:1–27, (2013). DOI: 10.1007/s10762-012-9946-2. x

[17] P. Kühne, C. M. Herzinger, M. Schubert, J. A. Woollam, and T. Hofmann, Invited article: An integrated mid-infrared, far-infrared, and terahertz optical hall effect instrument, Review of Scientific Intruments, 85, (2014). x

[18] L. Carreto, R. F. Madrigal, A. Fimia, et al., Study of angular responses of mixed amplitude—phase holographic gratings: Shifted Borrmann effect, *Opt. Lett.*, 26(11):786–788, (2001). x

[19] H. Kogelnik, Coupled wave theory of thick hologram gratings, *Bell Syst. Techn. J.*, 48(9):2909–2947, (1969). DOI: 10.1002/j.1538-7305.1969.tb01198.x. x

[20] J. B. Pendry, L. Martin-Moreno, and F. J. Garcia-Vidal, Mimicking surface plasmons with structured surfaces, *Science*, 305:847–848, (2004). DOI: 10.1126/science.1098999. x, xii

[21] F. J. Garcia-Vidal, L. Martin-Moreno, and J. B. Pendry, Surfaces with holes in them: New plasmonic metamaterials, *J. Opt. A-Pure Appl. Op.*, 7:S97, (2005). DOI: 10.1088/1464-4258/7/2/013. x, 5, 8

[22] T. Jiang, L. Shen, X. Zhang, and L.-X. Ran, High-order modes of spoof surface Plasmon polaritons on periodically corrugated metal surfaces, *Progr. Electromagn. Res. M*, 8:91–102, (2009). DOI: 10.2528/pierm09062901. x

[23] T. Gric, M. S. Wartak, M. Cada, J. J. Wood, O. Hess, and J. Pistora, Spoof plasmons in corrugated semiconductors, *J. Electromagnet. Wave*, 29:1899–1907, (2015). DOI: 10.1080/09205071.2015.1065772. x, xii, 29

[24] A. Rusina, M. Durach, and M. I. Stockman, Theory of spoof plasmons in real metals, *Appl. Phys. A*, 100(2):375–378, (2010). DOI: 10.1007/s00339-010-5866-y. x, 2, 3, 4, 43

[25] E. Ozbay, Plasmonics: Merging photonics and electronics at nanoscale dimensions, *Science*, 311:189–193, (2006). DOI: 10.1126/science.1114849. x

[26] W. L. Barnes, A. Dereux, and T. W. Ebbesen, Surface plasmon subwavelength optics, *Nature*, 424:824–830, (2003). DOI: 10.1038/nature01937. x

[27] W. Nomura, M. Ohtsu, and T. Yatsui, Nanodot coupler with a surface plasmon polariton condenser for optical far/near-field conversion, *Appl. Phys. Lett.*, 86:181108, (2005). DOI: 10.1063/1.1920419. x

[28] M. Quinten, A. Leitner, J. R. Krenn, and F. R. Ausscnegg, Electromagnetic energy transport via linear chains of silver nanoparticles, *Opt. Lett.*, 23:1331–1333, (1998). DOI: 10.1364/ol.23.001331. x

[29] R. Charbonneau, P. Berini, E. Berolo, and E. Lisicka-Shrzek, Experimental observation plasmon-polariton waves supported by a thin metal film of finite with, *Opt. Lett.*, 25:844–846, (2000). DOI: 10.1364/OL.25.000844. x

[30] B. Lamprecht, J. R. Krenn, G. Schider, H. Ditlbacher, M. Salerno, N. Felidj, A. Leitner, F. R. Aussenegg, and J. C. Weeber, Surface plasmon propagation in microscale metal stripes, *Appl. Phys. Lett.*, 79:51, (2001). DOI: 10.1063/1.1380236. x, xi

[31] T. Nikolajsen, K. Leosson, I. Salakhutdinov, and S. I. Bozhevolnyi, Polymer-based surface-plasmon-polariton stripe waveguides at telecommunication wavelengths, *Appl. Phys. Lett.*, 82:668, (2003). DOI: 10.1063/1.1542944. x

[32] J. R. Krenn, B. Lamprecht, H. Ditlbacher, G. Schider, M. Salerno, A. Leitner, and F. R. Aussenegg, Non-diffraction-limited light transport by gold nanowires, *Europhys. Lett.*, 60:663–669, (2002). DOI: 10.1209/epl/i2002-00360-9. x

[33] J. R. Krenn and J. C. Weeber, Surface plasmon polaritons in metal stripes and wires, *Philos. T. Roy. Soc. A*, 362:739–756, (2004). DOI: 10.1098/rsta.2003.1344. x

[34] S. A. Maier, P. G. Kik, H. A. Atwater, S. Meltzer, E. Harel, B. E. Koel, and A. A. G. Requicha, Local detection of electromagnetic energy transport below the diffraction limit in metal nanoparticle plasmon waveguides, *Nat. Mater.*, 2:229–232, (2003). DOI: 10.1038/nmat852. x, xi

[35] W. A. Murray, S. Astilean, and W. L. Barnes, Transition from localized surface plasmon resonance to extended surface plasmon-polariton as metallic nanoparticles merge to form a periodic hole array, *Phys. Rev. B*, 69:165407, (2004). DOI: 10.1103/physrevb.69.165407. x

[36] S. A. Maier, P. E. Barclay, T. J. Johnson, M. D. Friedman, and O. Painter, Low-loss fiber accessible plasmon waveguide for planar energy guiding and sensing, *Appl. Phys. Lett.*, 84:3990, (2004). DOI: 10.1063/1.1753060. x, xi

[37] S. A. Maier, M. D. Friedman, P. E. Barclay, and O. Painter, Experimental demonstration of fiber-accessible metal nanoparticle plasmon waveguides for planar energy guiding and sensing, *Appl. Phys. Lett.*, 86:071103, (2005). DOI: 10.1063/1.1862340. x

[38] P. Berini, R. Charbonneau, N. Lahoud, and G. Mattiussi, Characterization of long-range surface-plasmon-polariton waveguides, *J. Appl. Phys.*, 98:043109, (2005). DOI: 10.1063/1.2008385. x

[39] D. F. P. Pile and D. K. Gramotnev, Channel plasmon-polariton in a triangular groove on a metal surface, *Opt. Lett.*, 29:1069–1071, (2004). DOI: 10.1364/ol.29.001069. x

[40] S. I. Bozhevolnyi, V. S. Volkov, E. Devaux, J. Y. Laluet, and T. W. Ebbesen, Channel plasmon subwavelength waveguide components including interferometers and ring resonators, *Nature*, 440:508–511, (2006). DOI: 10.1038/nature04594. x, xi

[41] J. Chen, G. Smolyakov, S. R. J. Brueck, and K. J. Malloy, Surface plasmon modes of finite, planar, metal-insulator-metal plasmonic waveguides, *Opt. Express*, 16:14902–14909, (2008). DOI: 10.1364/oe.16.014902. x

[42] Y. Kurokawa and H. T. Miyazaki, Metal-insulator-metal plasmon nanocavities: Analysis of optical properties, *Phys. Rev. B* 75:035411, (2007). DOI: 10.1103/physrevb.75.035411. x

[43] H. Yang and J. Li, A highly efficient surface plasmon polaritons excitation achieved with a metal-coupled metal-insulator-metal waveguide, *AIP Adv.*, 4:127114, (2014). DOI: 10.1063/1.4903775. x

[44] G. Veronis and S. Fan, Bends and splitters in subwavelength metal-dielectric-metal plasmonic waveguides, *Appl. Phys. Lett.*, 87:131102-1–131102-3, (2005). DOI: 10.1063/1.2056594. x

[45] R. J. Walters, R. V. A. van Loon, I. Brunets, J. Schmitz, and A. Polman, A silicon-based electrical source of surface plasmon polaritons, *Nat. Mater.*, 9:21–25, (2010). DOI: 10.1038/nmat2595. x

[46] A. Lambrecht and V. Marachevsky, Casimir interaction of dielectric gratings, *Phys. Rev. Lett.*, 101:160403, (2008). DOI: 10.1103/physrevlett.101.160403. x

[47] G. Bimonte, Scattering approach to Casimir forces and radiative heat transfer for nanostructured surfaces out of thermal equilibrium, *Phys. Rev. A*, 80:042102, (2009). DOI: 10.1103/physreva.80.042102. x

[48] P. S. Davids, F. Intravaia, F. D. S. S. Rosa, and D. Dalvit, Modal approach to Casimir forces in periodic structures, *Phys. Rev. A*, 82:062111, (2010). DOI: 10.1103/physreva.82.062111. x

[49] R. Guerout, J. Lussange, F. S. S. Rosa, J.-P. Hugonin, D. A. R. Dalvit, J.-J. Greffet, A. Lambrecht, and S. Reynaud, Enhanced radiative heat transfer between nanostructured gold plates, *Phys. Rev. B*, 85:180301, (2012). DOI: 10.1103/PhysRevB.85.180301. x

[50] J. Lussange, R. Guerout, F. S. S. Rosa, J.-J. Greffet, A. Lambrecht, and S. Reynaud, Radiative heat transfer between two dielectric nanogratings in the scattering approach, *Phys. Rev. B*, 86:085432, (2012). DOI: 10.1103/physrevb.86.085432. x

[51] S. I. Bozhevolnyi, V. S. Volkov, E. Devaux, and T. W. Ebbesen, Channel plasmon-polariton guiding by subwavelength metal grooves, *Phys. Rev. Lett.*, 95:046802, (2005). DOI: 10.1103/physrevlett.95.046802. xi, xii

[52] A. V. Krasavin and N. I. Zheludev, Active plasmonics: Controlling signals in Au/Ga waveguide using nanoscale structural transformations, *Appl. Phys. Lett.*, 84:1416–1418, (2004). DOI: 10.1063/1.1650904. xi

[53] A. V. Krasavin, A. V. Zayats, and N. I. Zheludev, Active control of surface plasmon—polariton waves, *J. Opt. A: Pure Appl. Opt.*, 7:S85, (2005). DOI: 10.1088/1464-4258/7/2/011. xi

[54] P. Andrew and W. L. Barnes, Energy transfer across a metal film mediated by surface plasmon polaritons, *Science*, 306:1002–1005, (2004). DOI: 10.1126/science.1102992. xi

[55] P. Berini, Long-range surface plasmon polaritons, *Adv. Opt. Photon.*, 1:484–588, (2009). DOI: 10.1364/AOP.1.000484. xi

[56] K. Wang and D. M. Mittleman, Metal wires for terahertz wave guiding, *Nature*, 432(18):376–379, (2004). DOI: 10.1038/nature03040. xi

[57] G. Kumar, S. Pandey, A. Cui, and A. Nahata, Planar plasmonic terahertz waveguides based on periodically corrugated metal films, *New J. Phys.*, 13:033024, (2011). DOI: 10.1088/1367-2630/13/3/033024. xi

[58] L. Shen, X. Chen, and T.-J. Yang, Terahertz surface plasmon polaritons on periodically corrugated metal surfaces, *Opt. Express*, 16(5):3326–3333, (2008). DOI: 10.1364/oe.16.003326. xi

[59] S. A. Maier, S. R. Andrews, L. Martín-Moreno, and F. J. García-Vidal, Terahertz surface plasmon-polariton propagation and focusing on periodically corrugated metal wires, *Phys. Rev. Lett.*, 97(11):176805, (2006). DOI: 10.1103/physrevlett.97.176805. xi

[60] J. J. Wood, L. A. Tomlinson, O. Hess, S. A. Maier, and A. I. Fernandez-Dominguez, Spoof plasmon polaritons in slanted geometries, *Phys. Rev, B*, 85:075441, (2012). DOI: 10.1103/physrevb.85.075441. xi, xii

[61] G. Goubau, Surface waves and their application to transmission lines, *J. Appl. Phys.*, 21:1119–1128, (1950). DOI: 10.1063/1.1699553. xi, xii

[62] R. Ulrich and M. Tracke, Submillimiter waveguiding on periodic metal structure, *Appl. Phys. Lett.*, 22(5):251–253, (1973). DOI: 10.1063/1.1654628. xi, xii

[63] D. L. Mills and A. A. Maradudin, Surface corrugation and surface-polariton binding in the infrared frequency range, *Phys. Rev. B*, 39(3):1569–1574, (1989). DOI: 10.1103/physrevb.39.1569. xi, xii

[64] S. H. Mousavi, A. B. Khanikaev, B. Neuner III, Y. Avitzour, D. Korobkin, G. Ferro, and G. Shvets, Highly confined hybrid spoof surface plasmons in ultrathin metal-dielectric heterostructures, *Phys. Rev. Lett.*, 105:176803, (2010). DOI: 10.1103/physrevlett.105.176803. xi, xii

[65] D. Martin-Cano, O. Quevedo-Teruel, E. Moreno, L. Martín-Moreno, and F. J. García-Vidal, Waveguided spoof surface plasmons with deep-subwavelength lateral confinement, *Org. Lett.*, 36(23):4635–4637, (2011). DOI: 10.1364/ol.36.004635. xi

[66] A. J. Babadjanyan, N. L. Margaryan, and K. V. Nerkararyan, Superfocusing of surface polaritons in the conical structure, *J. Appl. Phys.*, 87:3785–3788, (2000). DOI: 10.1063/1.372414. 2, 5, 10, 48, 49, 50

[67] M. I. Stockman, Nanofocusing of optical energy in tapered plasmonic waveguides, *Phys. Rev. Lett.*, 93:137404, (2011). DOI: 10.1103/physrevlett.93.137404. 3

[68] B. Ferguson and X.-C. Zhang, Materials for terahertz science and technology, *Nat. Mater.*, 1:26–33, (2002). DOI: 10.1038/nmat708. 1, 36, 47

[69] M. Nagel, P. Haring Bolivar, M. Brucherseifer, H. Kurz, A. Bosserhoff, and R. Büttner, Integrated THz technology for label-free genetic diagnostics, *Appl. Phys. Lett.*, 80:154, (2002). DOI: 10.1063/1.1428619. 4, 5

[70] A. P. Hibbins, B. R. Evans, and J. R. Sambles, Experimental verification of designer surface plasmons, *Science*, 308:670–672, (2005). DOI: 10.1126/science.1109043. 5

[71] A. Rusina, M. Durach, K. A. Nelson, and M. I. Stockman, Nanoconcentration of terahertz radiation in plasmonic waveguides, *Opt. Express*, 16:18576–18589, (2008). DOI: 10.1364/oe.16.018576.

[72] M. Born and E. Wolf, *Principles of Optics*, University Press, Cambridge, 1999. DOI: 10.1017/9781108769914.

[73] J. Jung and T. G. Pedersen, Analysis of plasmonic properties of heavily doped semiconductors using full band structure calculations, *J. Appl. Phys.*, 113:114904, (2013). DOI: 10.1063/1.4795339.

[74] F. J. G. de Abajo and J. J. Sáenz, Electromagnetic surface modes in structured perfect-conductor surfaces, *Phys. Rev. Lett.*, 95:233901, (2005). DOI: 10.1103/PhysRevLett.95.233901. 1

[75] Z. Ruan and M. Qiu, Slow electromagnetic wave guided in subwavelength region along one-dimensional periodically structured metal surface, *Appl. Phys. Lett.*, 90:201906, (2007). DOI: 10.1063/1.2740174. 1

[76] S. A. Maier, S. R. Andrews, L. Martin-Moreno, F. J. Garcia-Vidal, Terahertz surface plasmon-polariton propagation and focusing on periodically corrugated metal wires, *Phys. Rev. Lett.*, 97:176805-1-4, (2006). DOI: 10.1103/physrevlett.97.176805. 1

[77] Y. Chen, Z. Song, Y. Li, M. Hu, Q. Xing, Z. Zhang, L. Chai, and C. Wang, Effective surface plasmon polaritons on the metal wire with arrays of subwavelength grooves, *Opt. Express*, 14:13021–13029, (2006). DOI: 10.1364/oe.14.013021. 8, 42, 43

[78] S. Q. Lou, T. Y. Guo, H. Fang, H. L. Li, and S. S. Jian, A new type of terahertz waveguides, *Chin. Phys. Lett.*, 23:235–238, (2006). DOI: 10.1088/0256-307x/23/1/068. 14

[79] L. M. Brehovskih, *Waves in Layered Media*, Academic Press, London, NY, 1960. 15, 23

[80] A. N. Lagarkov, S. M. Matitsin, K. N. Rozanov, and A. K. Sarychev, Dielectric properties of fiber-filled composites, *J. Appl. Phys.*, 84(7):3806–3818, (1998). DOI: 10.1063/1.368559. 15, 16

[81] O. Reynet, A.-L. Adenot, S. Deprot, and O. Acher, Effect of the magnetic properties of the inclusions on the high-frequency dielectric response of diluted composites, *Phys. Rev. B*, 66:094412, (2002). DOI: 10.1103/physrevb.66.094412. 15

[82] A. N. Lagarkov, A. K. Sarychev, T. R. Smychkovich, and A. P. Vinogradov, Effective medium theory for microwave dielectric constant and magnetic permeability of conducting stick composites, *J. Electromagn. Waves Applic.*, 6(9):1159–1176, (1992). DOI: 10.1163/156939392x00661. 16

[83] D. P. Makhnovskiy and L. V. Panina, Field-dependent surface impedance tensor in amorphous wires with two types of magnetic anisotropy: Helical and circumferential, *Phys. Rev. B*, 63:144424, (2001). DOI: 10.1103/physrevb.63.144424. 16, 20

[84] N. A. Usov, A. S. Antonov, and A. N. Lagarkov, Theory of giant magneto-impedance effect in amorphous wires with different types of magnetic anisotropy, *JMMM*, 185:159–173, (1998). DOI: 10.1016/s0304-8853(97)01148-7. 16

[85] M. Vazques and A. Hernando, A soft magnetic wire for sensor applications, *Phys. D, Appl. Phys.*, 29:939–951, (1996). DOI: 10.1088/0022-3727/29/4/001. 17, 23

[86] S. A. Baranov, Permeability of an amorphous microwire in the microwave band, *J. Communic. Techn. Electron.*, 48(2):226–228, (2003). 17

[87] V. N. Berzhansky, V. I. Ponomarenko, V. V. Popov, and A. V. Torkunov, Measuring the impedance of magnetic microwires in a rectangular waveguide, *Techn. Phys. Lett.*, 31(11):959, (2005). DOI: 10.1134/1.2136964. 17

[88] L. G. C. Melo, P. Ciureanu, and A. Yelon, Permeability deduced from impedance measurements at microwave frequencies, *JMMM*, 249:337–341, (2002). DOI: 10.1016/s0304-8853(02)00555-3. 17

[89] S. N. Starostenko, K. N. Rozanov, and A. V. Osipov, Microwave properties of composites with chromium dioxide, *JMMM*, 300:70–73, (2006). DOI: 10.1016/j.jmmm.2005.10.151. 20

[90] D. P. Makhnovskiy and L. V. Panina, Field dependent permittivity of composite materials containing ferromagnetic wires, *J. Appl. Phys.*, 93(7):4120–4129, (2003). DOI: 10.1063/1.1557780. 20, 21

[91] O. Acher, M. Ledieu, A.-L. Adenot, and O. Reynet, Microwave properties of diluted composites made of magnetic wires with giant magneto-impedance effect, *IEEE Trans. Magn.*, 39(5):3085–3090, (2003). DOI: 10.1109/tmag.2003.816011. 22, 24, 25

[92] S. N. Starostenko, K. N. Rozanov, and A. V. Osipov, Microwave properties of composites with glass coated amorphous magnetic microwires, *MISM, Book Abstr.*, 314, (2002). DOI: 10.1016/j.jmmm.2005.03.004. 23

[93] S. N. Starostenko, K. N. Rozanov, and A. V. Osipov, Microwave properties of composites with glass coated amorphous magnetic microwires, *JMMM*, 298:56–64, (2006). DOI: 10.1016/j.jmmm.2005.03.004. 23

[94] S. N. Starostenko and K. N. Rozanov, Microwave screen with magnetically controlled attenuation, *Pr. Electromagn. Res.*, 99:405–426, (2009). DOI: 10.2528/pier09060403. 25

[95] H. Seidel, L. Csepregi, A. Heuberger, and H. Baumgartel, Anisotropic etching of crystalline silicon in alkaline solutions, *J. Electrochem. Soc.*, 137:3612–3626, (1990). 25

[96] S. Li, M. M. Jadidi, T. E. Murphy, and G. Kumar, Terahertz surface plasmon polaritons on a semiconductor surface structured with periodic v-grooves, *Opt. Express*, 21:7041–7049, (2013). DOI: 10.1364/oe.21.007041. 29, 32

[97] G. Kumar, S. Li, M. M. Jadidi, and T. E. Murphy, Terahertz surface plasmon waveguide based on a one-dimensional array of silicon pillars, *New J. Phys.*, 15:085031, (2013). DOI: 10.1088/1367-2630/15/8/085031. 29

[98] L. Ding, W. Xu, C. Zhao, S. Wang, and H. Liu, Coupling of plasmon and photon modes in a graphene-based multilayer structure, *Optics Lett.*, 40:4524–4527, (2015). DOI: 10.1364/ol.40.004524.

[99] A. A. Maradudin, J. R. Sambles, and W. L. Barnes, *Modern Plasmonics*, vol. 4, Elsevier, (2014).

[100] A. A. Maradudin, I. Simonsen, J. Polanco, and R. M. Fitzgerald, Rayleigh and Wood anomalies in the diffraction of light from a perfectly conducting reflection grating, *J. Optics*, 18:024004, (2016). DOI: 10.1088/2040-8978/18/2/024004. 29

[101] P. A. D. Gonçalves and N. M. R. Peres, *An Introduction to Graphene Plasmonics*, World Scientific, (2016). DOI: 10.1142/9948. 30

[102] N. D. Mermin, Lindhard dielectric function in the relaxation-time approximation, *Phys. Rev. B*, 1:2362, (1970). DOI: 10.1103/physrevb.1.2362. 30

[103] A. Woessner, M. Lundeberg, Y. Gao, A. Principi, P. Alonso-Gonzlez, M. Carrega, K. Watanabe, T. Taniguchi, G. Vignale, M. Polini, J. Hone, R. Hillenbrand, and F. H. L. Koppens, Highly conned low-loss plasmons in graphene-boron nitride heterostructures, *Nat. Mater.*, 14:421–425, (2015). DOI: 10.1038/nmat4169. 30

[104] J. Kischkat, S. Peters, B. Gruska, M. Semtsiv, M. Chashnikova, M. Klinkmüller, O. Fedosenko, S. Machulik, A. Aleksandrova, G. Monastyrskyi, Y. Flores, and W. T. Masselink, Mid-infrared optical properties of thin films of aluminum oxide, titanium dioxide, silicon dioxide, aluminum nitride, and silicon nitride, *Appl. Optics*, 51:6789–6798, (2012). DOI: 10.1364/ao.51.006789. 30

[105] R. Li, C. DAgostino, J. McGregor, M. D. Mantle, J. A. Zeitler, and L. F. Gladden, Mesoscopic structuring and dynamics of alcohol/water solutions probed by terahertz time-domain spectroscopy and pulsed field gradient nuclear magnetic resonance, *J. Phys. Chem. B*, 118:10156–10166, (2014). DOI: 10.1021/jp502799x. 30

[106] B. Ng, J. Wu, S. M. Hanham, A. I. Fernandez-Domnguez, N. Klein, Y. F. Liew, M. B. H. Breese, M. Hong, and S. A. Maier, Spoof plasmon surfaces: A novel platform for THz sensing, *Adv. Optical Mater.*, 1:543–548, (2013). DOI: 10.1002/adom.201300146. 34

[107] D. K. Efetov and P. Kim, Controlling electron-phonon interactions in graphene at ultrahigh carrier densities, *Phys. Rev. Lett.*, 105:256805, (2010). DOI: 10.1103/physrevlett.105.256805. 34

[108] J. Ye, M. F. Craciun, M. Koshino, S. Russo, S. Inoue, H. Yuan, H. Shimotani, A. F. Morpurgo, and Y. Iwasa, Accessing the transport properties of graphene and its multi-layers at high carrier density, *Proc. of the National Academy of Sciences*, 108:13002–13006, (2011). DOI: 10.1073/pnas.1018388108. 35

[109] C. R. Dean, A. F. Young, I. Meric, C. Lee, L. Wang, S. Sorgenfrei, K. T. Watanabe, P. K. Taniguchi, K. L. Shepard, and J. Hone, Boron nitride substrates for high-quality graphene electronics, *Nat. Nanotechnol.*, 5:722–726, (2010). DOI: 10.1038/nnano.2010.172. 36

[110] J. J. Foley IV, H. Harutyunyan, D. Rosenmann, R. Divan, G. P. Wiederrecht, and S. K. Gray, When are surface plasmon polaritons excited in the Kretschmann–Raether config-uration?, *Sci. Rep.*, 5:9929, (2015). DOI: 10.1038/srep09929. 36

[111] A. Otto, Excitation of nonradiative surface plasma waves in silver by the method of frustrated total reflection, *Zeitschrift für Physik*, 216:398–410, (1968). DOI: 10.1007/bf01391532. 36

[112] T. R. Zhan, F. Y. Zhao, X. H. Hu, X. H. Liu, and J. Zi, Band structure of plasmons and optical absorption enhancement in graphene on subwavelength dielectric gratings at in-frared frequencies, *Phys. Rev. B*, 86:165416, (2012). DOI: 10.1103/physrevb.86.165416. 36, 37

[113] I. Khromova, A. Andryieuski, and A. Lavrinenko, Ultrasensitive terahertz/infrared waveguide modulators based on multilayer graphene metamaterials, *Laser Photon. Rev.*, 8(6):916–923, (2014). DOI: 10.1002/lpor.201400075. 36

[114] V. M. Agranovich and V. E. Kravtsov, Notes on crystal optics of superlattices, *Solid State Commun.*, 55(1):85–90, (1985). DOI: 10.1016/0038-1098(85)91111-1.

[115] I. Iorsh, A. Orlov, P. Belov, and Y. Kivshar, Interface modes in nanostructured metal-dielectric metamaterials, *Appl. Phys. Lett.*, 99:151914, (2011). DOI: 10.1063/1.3643152. 36

[116] L. A. Falkovsky, Optical properties of graphene, *J. Phys. Conf. Ser.*, 129:012004, (2008). DOI: 10.1088/1742-6596/129/1/012004. 36

[117] G. W. Hanson, Dyadic Green's functions and guided surface waves for a surface conductivity model of graphene, *J. Appl. Phys.*, 103(6):064302, (2008). DOI: 10.1063/1.2891452. 42, 43

[118] A. Vakil and N. Engheta, Transformation optics using graphene, *Science*, 332(6035):1291–1294, (2011). DOI: 10.1126/science.1202691. 43

[119] Y.-C. Chang, C.-H. Liu, C.-H. Liu, S. Zhang, S. R. Marder, E. E. Narimanov, Z. Zhong, and T. B. Norris, Realization of mid-infrared graphene hyperbolic metamaterials, *Nat. Commun.*, 7:10568, (2016). DOI: 10.1038/ncomms10568. 43

[120] J. S. Gomez-Diaz, C. Moldovan, S. Capdevila, J. Romeu, L. S. Bernard, A. Magrez, A. M. Ionescu, and J. Perruisseau-Carrier, Self-biased reconfigurable graphene stacks for terahertz plasmonics, *Nat. Commun.*, 6:6334, (2015). DOI: 10.1038/ncomms7334. 43

[121] T. Gric and O. Hess, Controlling hybrid-polarization surface plasmon polaritons in dielectric-transparent conducting oxides metamaterials via their effective properties, *J. Appl. Phys.*, 122:193105, (2017). DOI: 10.1063/1.5001167. 45, 46

[122] P. B. Johnson and R. W. Christy, Optical constants of the noble metals, *Phys. Rev. B*, 6:4370, (1972). DOI: 10.1103/physrevb.6.4370. 43, 45, 46

[123] D. Woolf, M. Kats, and F. Capasso, Spoof surface plasmon waveguide forces, *Opt. Lett.*, 39:517–520, (2014). DOI: 10.1364/ol.39.000517. 43

[124] N. Yu, Q. J. Wang, M. A. Kats, J. A. Fan, S. P. Khanna, L. Li, A. G. Davies, E. H. Linfield, and F. Capasso, Designer spoof surface plasmon structures collimate terahertz laser beams, *Nat. Mater.*, 9:730–735, (2010). DOI: 10.1038/nmat2822. 49

[125] C. R. Williams, S. R. Andrews, S. A. Maier, A. I. Fernandez-Dominguez, L. Martin-Moreno, and F. J. Garcia-Vidal, Highly confined guiding of terahertz surface plasmon polaritons on structured metal surfaces, *Nat. Photonics*, 2:175–179, (2008). DOI: 10.1038/nphoton.2007.301. 54

[126] Y.-Q. Liu, L.-B. Kong, and P.-K. Liu, Long-range spoof surface plasmons on the doubly corrugated metal surfaces, *Opt. Commun.*, 370:13–17, (2016). DOI: 10.1016/j.optcom.2016.02.059.

66 BIBLIOGRAPHY

[127] T. Gric, Spoof plasmons in corrugated transparent conducting oxides, *J. Electromagnet. Wave*, 30:721–727, (2016). DOI: 10.1080/09205071.2016.1145076.

[128] G. V. Naik, V. M. Shalaev, and A. Boltasseva, Alternative plasmonic materials: Beyond gold and silver, *Adv. Mater.*, 25:3264–3294, (2013). DOI: 10.1002/adma.201205076.

[129] G. V. Naik, J. Kim, and A. Boltasseva, Oxides and nitrides as alternative plasmonic materials in the optical range, *Opt. Mater. Express.*, 1:1090–1099, (2011). DOI: 10.1364/ome.1.001090.

[130] E. Verhagen, M. Spasenovic, A. Polman, and L. Kuipers, Nanowire plasmon excitation by adiabatic mode transformation, *Phys. Rev. Lett.*, 102:203904, (2009). DOI: 10.1103/physrevlett.102.203904.

[131] R. S. Anwar, H. Ning, and L. Mao, Recent advancements in surface plasmon polaritons-plasmonics in subwavelength structures at microwave and terahertz regime, *Dig. Commun. Netw.*, 4(4):244–257, (2017). DOI: 10.1016/j.dcan.2017.08.004.

[132] E. J. C. Dias and N. M. R. Peres, Controlling spoof plasmons in a metal grating using graphene surface plasmons, *ACS Photonics*, 4(12):3071–3080, (2017). DOI: 10.1021/acsphotonics.7b00629. 14, 29, 31, 32, 33

Author's Biography

TATJANA GRIC

Dr. Gric is a researcher of outstanding potential. Her research career has been focused on the investigation of waveguide devices. Another major goal of her studies is plasmonics. Moreover, the broad scope of research carried out by Dr. Gric has included investigations into the new fascinating properties of novel materials. Dr. Gric also has a record of effective teaching in the rank of Associate Professor. Dr. Gric has published extensively in her field of investigation with more than 40 peer-reviewed papers in top journals in physics, electrodynamics, and optics. It is worth noting that her recent publication rate is getting even higher with her being the first author.

Printed in the United States
by Baker & Taylor Publisher Services